Man v Machine

Journey of Complacency

Man v Machine

Spiderwize
Remus House
Coltsfoot Drive
Woodston
Peterborough
PE2 9BF

www.spiderwize.com

A CIP catalogue record for this book is available from the British Library.

ISBN: 978-1-912694-01-3

Man v Machine

Journey of Complacency

Paul J Mahoney

SPIDERWIZE
Peterborough UK
2018

For family

past, present and future

Then out spake brave Horatius,
The Captain of the Gate:
"To every man upon this earth
Death cometh soon or late.
And how can man die better
Than facing fearful odds,
For the ashes of his fathers,
And the temples of his gods?
-From Thomas Babington Macaulay's
Lays of Ancient Rome

Man can die better by dying in bed with his family around him after a long and happy life and if he is lucky his last words will leave his family with something profound! That they can carry forward and pass on to their families. Many people in my situation would get angry and forever be twisted about their new life. I always wanted people to learn from my mistake and that of others.

And that's why you can't get angry; you have to get smart.
– Vernon Jordan

Contents

Introduction

So, why write a book? It is normally, because someone wants to part knowledge or tell a story (true or fictional).

My book is going to talk about both. The events that lead to myself becoming disabled, but the luckiest man alive and what I have learnt since the 25th November 2000 at around 8am on that Saturday morning. The book is an extension of the presentations that I have run all over the globe to countless people to ensure they and their employees go home every day in one piece.

I'm not going to pull any punches in this book, because it is how I see the world. I will warn you as it is my duty of care to do so that some topics will upset in places. It is real life and while I write this book someone somewhere will have experienced the pain and hurt of being injured at work. Actually it won't be just one person, it will be many and it is not just them that suffer. It will be families, friends and colleagues that will be going through the pain and hurt.

The question that will unite them all.

WHY AND HOW COULD THIS HAVE HAPPENED?

The aim of this book is to get yourself to take the Road to Damascus – where you have that sudden turning point, where you need to say to yourself and others, we need to take care, we need to change!

I hope this story about Paul O'Neill kick starts that journey and inspiration.

A few minutes before noon, the new chief executive, Paul O'Neill, took the stage. He looked dignified, solid. Confident. Like a chief executive.

Then he opened his mouth.

"I want to talk to you about worker safety," he said. "Every year, numerous Alcoa workers are injured so badly that they miss a day of work.

"I intend to make Alcoa the safest company in America. I intend to go for zero injuries."

The audience was confused. Usually, new CEOs talked about profit margins, new markets and 'synergy' or 'co-opetition.' But O'Neill hadn't said anything about profits. He didn't mention any business buzzwords at all in fact.

Eventually, someone raised a hand and asked about inventories in the aerospace division. Another asked about the company's capital ratios.

"I'm not certain you heard me," O'Neill said. "If you want to understand how Alcoa is doing, you need to look at our workplace safety figures." Profits, he said, didn't matter as much as safety.

The investors in the room almost stampeded out the doors when the presentation ended. One jogged to the lobby, found a pay phone, and called his 20 largest clients.

"I told them, I said, 'The board put a crazy hippie in charge and he's going to kill the company,'" that investor told me. "I ordered them to sell their stock immediately, before everyone else in the room started calling their clients and telling them the same thing.

How wrong that was.

Within a year of O'Neill's speech, Alcoa's profits would hit a record high. By the time O'Neill retired in 2000 to become Treasury Secretary, the company's annual net income was five times larger than before he arrived, and its market capitalization had risen by $27 billion. Someone who invested a million dollars in Alcoa on the day O'Neill was hired would have earned another million dollars in dividends while he headed the company, and the value of their stock would be five times bigger when he left.

What's more, all that growth occurred while Alcoa became one of the safest companies in the world.

What Paul O'Neill highlighted was that profits will come, but it's not all about the numbers and bottom lines. What Paul O'Neill hit on was that the less a company spends on investigations, staff sickness and downtime of machinery, the more the company makes. Profits rise, as the people employed to make the products are where they should be, working on the machinery producing or delivering the projects being built.

So, where do I start to tell this everyday tale of man vs machine and the journey of complacency? The only sensible starting point is about 7am on the 25th November 2000.

25th November 2000

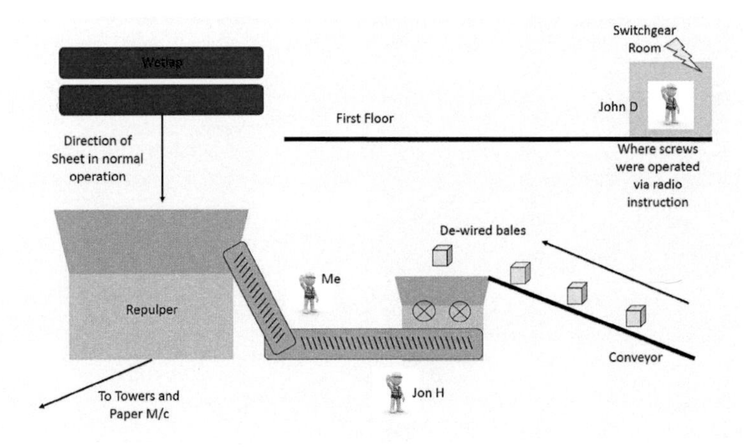

Fig: 1 Wetlap Layout

I was called to say that the paper machine next door was ramping up for 100% recycled pulp.

My colleague (we'll call him John D), started the motors in the switchgear room on the first floor.

I fed the first bales in, which were sitting on the infeed conveyor from the previous run a couple of weeks beforehand. The first blockage occurs about five minutes later. Another colleague (we'll call John H) helped dig out the blocked pulp at the crossover point. After forty minutes the blockage is cleared and the machine grinds back into life.

All was fine for a minute or two when it annoyingly blocks again. We decide to repeat the process. But time is money as they say, so half way through John H and I decide to run a screw backwards to help speed up proceedings. Then, to stop a third blockage, both John H and I agree that we need to check the crossover point. Due to the pulp going through the screw conveyors is dry this was causing the issues.

John H radios John D to stop so I can check the crossover point. This meant John D could then start again after the check.

Right at that point, John H radio fails. He keeps trying but for some reason it was completely dead. I can see the screw moving, so keep clear.

John H and I swap radios to save time, plus we're close enough to sign to each other anyway (I can get a new radio on way back to the Wetlap hut).

The screw stops. John H and I give each other the thumbs up to say all clear and I can start the check.

I put my arm through the hatch, and start to feel the screws to check that the crossover is clear of anymore pulp.

Suddenly at 1400RPM…

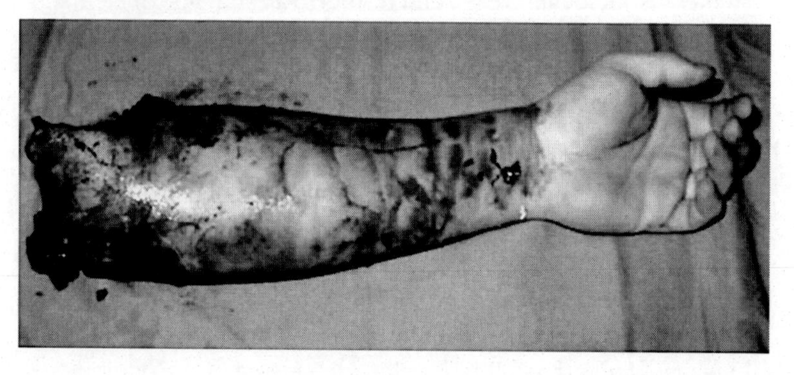

Fig 2: My arm

The reconstruction is different from what I said previously on the other page because the film was made during the legal cases. It gives an idea of the working environment that we faced around that particular piece of machinery.

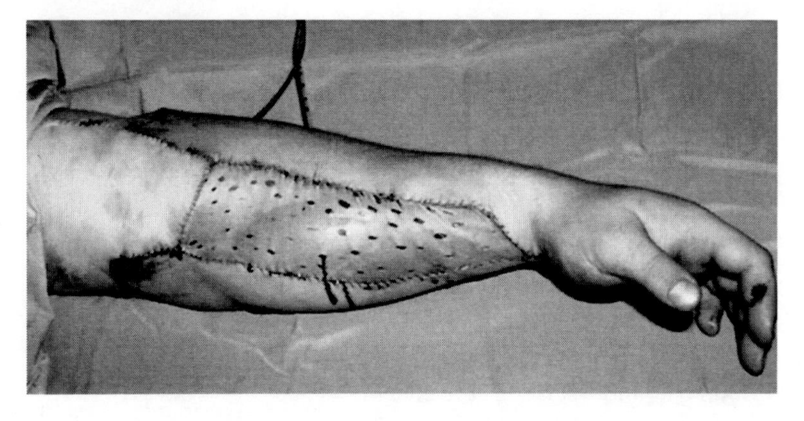

What you have just read and watched and looked at, is the end product of eighteen months in the making.

Accidents/incidents are the end results to a catalogue of elements that have evolved (and allowed) over time. Not just days, but months or even years.

These elements are not just evolving in one place, they are happening in various workplaces and departments, but no-one links them because they are treated as separate entities. When the major incident happens, you can see traces of the separate occurrences weaved throughout.

Incidents good and bad are the result of a journey. In the next few pages I am going to take you back to the start of the journey in my unique way. I want to highlight where the elements are evolving to potentially cause that major accident/incident in your workplace.

Safety. It's All Relative

Let's go back to the beginning with this simple question: Why do people at work do things that are dangerous or out of character?

The best way to sum things up is the quote from the film 10 Cloverfield Lane by the character Howard when he said 'People are strange creatures. You can't always convince them that ***safety*** is in their best interest'.

Why is it that you can't convince them of this? It is because of the sublime messages dissipated around the workplace. Targets, production, quality, the sound of machines running and financial figures all focus the mind to keep delivery on track.

There is another thing and it is to do with attitude to safety depending on the industry concerned. For example, a roofer or scaffolder aged over, say, 50 is less likely to consider health and safety issues as a matter of course than someone younger. The new generation of workers are in a more fortunate situation because the previous generations have learnt and built on the mistakes of the past.

Speaking of the older generation, when I joined the paper mill, I was paired up with the most experienced person there. He was going to teach me everything to do with my job and become my mentor. This is nothing unusual as most people are paired up this way. What is unusual is what I was told; one day my mentor held up his right hand and said as he wagged what he had left of his index finger. 'You'll never be a papermaker until you lose a digit!'

I nodded, like you do, and try to compute what has just been said. Thinking back now, and I won't know for sure, but was he covering up for his blushes or was he giving me the armour for the future for if I was to lose a finger, it would be my badge of honour.

Papermakers used to have a missing digit as a badge of honour. Other industries have theirs to. I had the honour to talk to a mix of trades working on the Palaces of Westminster and during the showing of the edited 90 second video of my accident, I heard one of the scaffolders say 'you wouldn't catch me doing that!'

I agreed and said you wouldn't catch me balancing thirty feet up on a beam with no protection. As the session went on I told the story about the missing digit. The scaffolder was about to say something about not having badges of honour, when I asked "Is it true that you're not a scaffolder till you have bounced from 15 feet?"

The scaffolder looked at me with his first fall written all over his face.

We need to remember that no single industry is any different from the next. The only difference is that people *think* they are different.

The Essence of Work

"Laziness may appear attractive, but work gives satisfaction." - Anne Frank

We need to look at the essence of what work actually is - or should I say the goal and vision of the organisation that we work for. It simply boils down to getting a product or service from point A to point Z, using the least capital or effort, but achieving maximum profit.

This is the same when it comes to any employee v boss relationship. The boss wants maximum output for minimum input and the employee wants minimum input for maximum pay.

Watch anybody at work, driving a car or just watching the TV. They will put in the minimal amount of effort, but expect maximum output.

Let us go back to the A to Z process, and use F1 as the analogy to cement the idea. With just one car on the track the effort to win is easy, as there are no opponents. The team maximises it chances by adding another car, just in case the other one breaks down.

Once other teams, cars and possibly better drivers are added, the A to Z process becomes harder to achieve. So the team adapts a different process to meet it targets.

This is where the testing of the rules starts to take place. If loop holes are found, then they are exploited to their potential until the loop hole or holes are closed.

This is where the Health and Safety at Work Act 1974 comes in. It places responsibility on organisations, employers and employees (we must not forget employees) to have legal, moral and financial accountabilities to ensure each keep to the law.

The trouble comes when you break the essence of the health and safety act down to the same analogy as before with F1. The Health and Safety Executive says you (racing team) need to get your car around the track (x number of laps) to win the race. The organisation (employer) needs to work out how to do it.

The Health and Safety Executive does help by supplying ACoPs (Approved Codes of Practice). These are not the gold standard, but are of a high enough standard to ensure good practice to comply with the law.

As with any set of rules, they are open to interpretation because we each see things differently. How many times have you watched a manager or the coach of a team being asked about an incident and they turned around and said I didn't see it or I need to see the replay or speak to the player?

What is happening is that they are buying time or just deflecting from the incident.

As mentioned, the essence of work is getting from A to Z, in the shortest way possible. This philosophy has been drummed into employees over a long period of time.

So, they arrive at work knowing they need to get raw material (A) to Delivery (Z). They have been trained to carry out the task and are aware of the health and safety. The organisation is content that the employees have the right SKATE (Skill, Knowledge, Ability, Training and Experience) to complete the task, which is fine while everything is going to plan (A, B, C, D >>>>>>>>> W, X, Y, Z).

Even if a step is missed (A, B, D, E >>>>>>>>> W, X, Y, Z) things are usually alright. It's when whole sections (A, E, G, J >>>>>>>> R, X, Y, Z) start to be missed or new sequences introduced (A, B, C, 1, 2, 3 >>>>>>> 7, 8, 9, Y, Z) when the trouble starts. This is because employees get away with it and the bosses either realise it and are happy as more is getting delivered at no extra cost, or, they are not aware because the communication channels have broken down.

The picture below sums up what I have written about. I took the photograph out walking my dog. The photograph is of the finished article of part of the bigger gas mains replacement happening near to where I live.

The quality of the gas mains replacement is not an issue, the issue is the finished look. The question that needs to be asked here is what element is missing in the sequence?

What was communicated and how was it understood?

Another question needs to be asked about organisational pride and personal pride.

Would you be happy to leave the job like this?

Fig 3: Bus Stop in Garlinge

The essence of work is setting the right standards and culture and ensuring it is followed through from A to Z. The wrong tone, word or action starts the ball rolling down the wrong path.

As a manager said to me once about a complaint I made about

a penis drawn on my car service paperwork: 'Boys will be boys in the workshop'. They wonder why I never went back to buy a new car or have the car MOT or serviced…

> *"Success is no accident. It is hard work, perseverance, learning, studying, sacrifice and most of all, love of what you are doing or learning to do".* - Pele

The Perfect Crime

Sometimes the hardest thing about committing the perfect crime can be keeping your genius to yourself. - Sarah Lacy

I was listening to Punt PI the other day. It is a radio series on Radio 4 about unsolved crimes and mysterious incidents.

This particular episode was about an army officer dying after eating a poisoned pheasant. One guest thinks it was the gravy that was poisoned and not the poor pheasant, as gravy is easier to poison.

With any crime there are suspects and with these suspects you have to look at their means, motive and opportunity in committing the crime.

Means are the tools used in committing the crime and were they available to the suspect.

Motive is the reason for committing the crime.

Opportunity is the time available to commit the crime.

If a suspect has all three elements to commit the crime, then more than likely they are the assailant or the prime suspect until solid proof is established.

Now bring these three elements of means, motive and opportunity into the workplace. Using these three elements we can potentially establish where short cutting (bypassing the system) will happen.

Means - what tools are around a piece of equipment. These tools can be in view or hidden but easy for the operative to get to in the event of breakdown.

Motive - what is the desire to keep things running and on schedule. Production plans and previous records play their part. Language used to get results (including body language).

Opportunity - has there been time in the past to develop systems that allow the operative/s to find a way to get the job done quicker.

Accidents/incidents are not one-off events (as in they are the first time someone tried to bypass the system). Elements are learned slowly and separately until there is an event that means, motive and opportunity show themselves to the operative/s in committing the crime.

99% of the time the crime is perfect as the operative/s use the 3 elements to their advantage in keeping things (production or delivery) on track. At times they even get congratulated in breaking records or getting things back on track.

This perfect crime continues not all the time, but when needed to keep production, delivery and quality of product at its best.

With how to commit the prefect crime, tweaked and learned by others to keep production, delivery and quality at its peak. 3 dangerous new factors emerge from the shadows.

Time - how much longer can the practice last before an incident or accident occurs.

Assumptions - the mindset of the operative/s that what they are doing is fine and no repercussions will happen to them or fellow workers if caught as the work is being completed

Satisfaction - operative/s are comfortable in operating in an unsafe manner, to get things sorted as the job is being rewarded.

Short cutting is a subtle thing that is not waving a big banner saying crime. It is in the shadows of operations to achieve the KPIs of a company. Most of the time, they are innocent little actions (forgetting to put the seatbelt on when on a forklift truck) and no one is hurt. There is through a time when people overreach the limit and bang something is damaged or worse people are hurt or killed.

In these times when injury non or fatal occur that the reset button is hit, and everyone takes the stance of judger or learner of the incident and those in involved.

To stop the prefect crime, we need to take a look at operations in the raw light of day when it's not happening like it should be happening and ask how to do we really do things around here and is there a consensus to achieve in sorting out the issue safely while still meeting the goal of the company.

It's a matter of everyone (operators to managers to ultimately directors) being honest about how operations happen around here and how we can achieve safe operations 100% of the time without damage or injury.

Environment

Complacency refers to the loss of the fear of injury that typically motivates employees to work safely.

Too often the problem is that employees become complacent and begin to shortcut safety procedures.

These two sentences have been taken from Terry McSween's book Value Based Safety Process and his chapter about complacency. I'm not going to disagree with Terry McSween because employees do lose fear in their job because they do it every day.

If you want to see complacency in action just step out of your front door and walk to the nearest bench. Watch the vehicles going up and down the road and then people walking on the pavement and crossing the road.

As you sit there and watch the world, you will see complacency in action. Drivers speed or cut across junctions and people crossing the road don't even look before they cross.

The question has to be asked - why? The answer is environment. People get used to the environment that they are operating in. With familiarity comes complacency.

I'm going to prove it with two battles, two sporting events and a businessman who cost his company a reported £500 million.

1066 - The Battle of Hastings. As the shield wall held against the Norman invasion, it was looking more likely that Harold was going to win the day. That is, until part of the shield wall chased the fleeing Norman ranks of one of the flanks down the hill. At the bottom, the Normans regrouped and cut down the Anglo Saxons. With the shield wall weakened the Norman's won the battle and changed the course of British and European history.

The Anglo Saxons got complacent as the Normans were starting to lose the battle. They got careless because they took the Normans' tactic as defeat. The Anglo Saxon's paid the price for not reading the environment.

Seven-hundred and fifty years later - the battle of Trafalgar was raging on the high seas. The ships from Britain and France waged all out destruction against each other. Nelson stood on the top deck of HMS Victory in all his stature of an admiral of the fleet. As musket balls flew all around, Nelson's underlings begged him to follow them and take off his medals as the French snipers was using them as targets. Nelson is said to have told them that the medals were won with pride and he was not taking them off!

Within minutes Nelson was below deck breathing his last breath, as a French sniper got lucky and shot him in the chest.

Nelson paid the price because he judged the environment as not involving him. He thought he was above the situation going on around him.

1968 – The Rugby League Challenge Cup Final boiled down to the last kick of the game. Wakefield Trinity had just scored a try and it was now time for Don Fox to kick the conversion and win the game. The weather was rotten and it had rained heavily throughout the game.

As Don stepped up he was soaked throughout; the ball was heavy with water. He had already kicked the points needed to keep Wakefield in the game. He was comfortable with the task ahead, but he had kicked the previous points when the ball was a little less soaked and time wasn't now on his side.

This was the last kick of the game. If you have played any rugby or football game, where the ball and yourself are soaked you try the best you can to dry ball, boot and top to get the cleanest of contacts. Don didn't do that, he just kicked the ball and sliced it wide, handing the trophy to Leeds.

Don got caught up in the environment of winning the trophy and having successfully kicked previously in the better conditions.

History and sport has illustrated complacency happens in the heat of battle. It could never happen to a successful business man like Gerald Ratner of Ratner's talking to his peers surely? Wrong. In 1991 he was talking to the Institute of Directors about his success. Assuming it was a closed conference, he basically - loudly - told his peers he sold crap. At the end of the hall were some journalists who had just heard and written down gold dust.

With that throw away comment, intended to be a joke, he cost a company hundreds of jobs and £500 million off the balance sheet. Not to mention shattering his reputation.

Ratner took it for granted the environment he was speaking at was just his peers. Why would journalists want to be at the IoD, he reasoned. He felt invincible.

Finally, we look at the Belgian Grand Prix of 2015. The most professional of sports, because everyone is so drilled in what they do. Mechanics can change a set of wheels in under two seconds.

So, when Bottas came into the pits what could go wrong, apart from a slow pit stop. The mechanics did their job and released Bottas back into the race, only to be told seconds later that he had to come back in for a time penalty. It wasn't for releasing him into traffic, but because the right rear was a different compound to the rest. The tyres had got mixed up and the right rear was odd. How do we know? The tyres have different colour stamps printed around them for fans at the race and at home to see what tyres the teams are using.

In the heat of the race, especially during the fast pit stop, the tyres were mixed and no-one noticed until it was too late. The team got complacent as they complete hundreds of tyre changes a year and the tyres are always the right compound. The environment of pressure to get the driver out in front of the pack took over and in the end it cost the Williams team constructor points and Bottas a podium.

Sunday Roast

It's Sunday morning. That is how I like to set the scene when telling a story like this one.

A young woman is preparing a pot roast while her friend looks on. She cuts off both ends of the roast, prepares it and puts it in the pan.

"Why do you cut off the ends?" her friend asks.

"I don't know", she replies. "My mother always did it that way and I learned how to cook it from her".

Her friend's question made her curious about her pot roast preparation. During her next visit home, she asked her mother,

"How do you cook a pot roast?" Her mother proceeded to explain and added, "You cut off both ends, prepare it and put it in the pot and then in the oven".

"Why do you cut off the ends?" the daughter asked. Baffled, the mother offered, "That's how my mother did it and I learned it from her!"

Her daughter's inquiry made the mother think more about the pot roast preparation. When she next visited her mother in the nursing home, she asked, "Mum, how do you cook a pot roast?" The mother slowly answered, thinking between sentences.

"Well, you prepare it with spices, cut off both ends and put it in the pot". The mother asked, "But why do you cut off the ends?" The grandmother's eyes sparkled as she remembered. "Well, the roasts were always bigger than the pot that we had back then. I had to cut off the ends to fit it in".

The question is why do you do what you do? It is because when we join work as youngsters (or even as we get older), we take things for granted. We accept that things are just done a certain way and that's that.

The mill I used to work in was nearly 100 years old. Whole families around Swale worked there, or had worked there, or will have future generations working there.

Status Quos (the state before the war) are developed amongst shifts, departments and sites. Being the newbie, we don't want to rock the boat, so we tow the party line.

You See, but Don't Observe

Airports take up a lot of my time. The number of people who sit on planes and don't watch either the flight attendants' drills, or the videos explaining the safety procedures is shocking.

So, if people are like this, how can you (the safety manager, if you are one) expect the guys or girls on the floor sit and watch safety videos or training courses?

Now, consider your organisation. What, is the safety culture like? And the follow up question: Why is it like that?

Now back to the plane. I understand some have a fear, some want to sleep, some fly all the time and some need to make that last call, text, email to work or home. But come on people, that five minutes could be the difference between seeing your family again or not.

A friend, who is a flight attendant, was telling me that after a plane crash recently, she had every eye in that plane watching her and her colleagues going through the safety demo. Why is it that a disaster has to happen (small or large) before people start to take note of the environment they're in? And more importantly, how long is it before things get comfortable again? What is the trigger that starts the slide?

The comedian Dave Allen summed it up beautifully in one of his stand-up routines. As he said, when I get on a plane all I'm interested in, is the following:

Will it take off?

Will it stay in the air?

Will it come down where it's expected?

Everybody in an Organisation is like this in some way!

Will I get to work on time?

Will I have an easy day?

Will I get home on time?

You can relate what he said to the organisation that you work in. Some people will rush to work due to being late or they just want to be there early (as they think the sooner they're at work the sooner they'll get home). Some are thinking about having an easy day as they don't feel well or others just want to get home (as it is a happier place). Or they need to get to the next meeting or delivery.

Understanding the World

How do we see the world? We see the world through our job titles or description because while we are awake that is where we spend most of our time at work. There are other things too, like how we are brought up and the friends we mix with, plus some other things too.

One thing that is overlooked is being left handed!

If you are left handed, there is one thing that you will have been called at some point, and that is cack-handed. Left-handed people that are always asked to do the difficult tasks, like to reach the awkward nuts and bolts.

Left handed people see the world completely differently because they have to almost sit on the fence as they judge the left-handed and right-handed way of doing things. Which way is easier?

The question that needs to be asked is how do you see rules?

Black and white?

Grey?

Multi-coloured?

If you can't decide, my advice is you can only sit on thefence so long before it gives you splinters.

To illustrate the point, watch the video.

For me, the ball was just in play as Law 9 of association football says that the whole ball must cross the line to be in or out of play.

You can understand why the player picked up the ball because he can see the green of the grass. It's a semi-final so the pressure is on to get the ball back to the keeper and back up to the other end and into the opponent's goal.

Bring that situation in to the workplace. Are risk assessments and safe systems of work either black and white, grey or multicoloured? The answer is down to you – your perception at that moment in time.

I'm going to paint two pictures for you. One is probably the greatest scene in sci-fi film history - the moment Darth Vadar tells Luke Skywalker that he is his father. Luke is hanging on to a plinth hundreds of metres up in the air. That is the film version - in reality Luke Skywalker is about 10 foot up in the air and below him are mattresses to protect him if he was to fall.

The other is one of a cricket game (test match, the 5 day version). The fielding team are either getting tired or they need to throw a new dimension to the game. The spin bowler is played and as he bowls his over, the fielders are all around the batman ready to catch the ball. The bowler spins the ball at some fantastic angles. Unless the batman hits the ball for a four or a six, every ball is a wicket, as bowler and fielders appeal as one.

Both scenes are open to your perception of how you see the world. Rules are no different as they are open to everyone's interpretation. Risk assessments and safe systems of work are no different because at work the job has to get done. So, rules are bent (not broken) to fit the situation.

Fig 4&5 Temporally fix, permanent solution

I'm lucky enough to have site tours of the

companies I talk to. Sometimes I ask if I can take photographs of what I see, to offer a fresh eye over what I witness.

I took the above photographs on a tour (I will not name the company). In these photos, protection devices are protecting electric eyes. One was made out of a cardboard tube and the other was a rubber pipe (hose) covered in tape and secured by a jubilee clip.

The photographs were used as part of what I saw on site and talked about later. When does temporary become permanent?

When shown to the guys and girls on the shop-floor, they said 'you need to speak to the Engineering Manager, he's on the next session'!

So, I asked the question, when does temporary become permanent? The Engineering Manager spoke up with a very defensive attitude and said 'there is nothing wrong with these, I can't see why you would pick these out! They are doing the job'!

'There is always an easy solution to every problem - neat, plausible, and wrong'.

H. L. Mencken

These photographs highlight people's shortcuts (or skewed view) in some way. Personally I don't like the word 'shortcut', as it is implies that someone is lazy. Actually people do what they do because they are trying to do one of the following:

1. Time - trying to save it
2. Money - trying to get that extra buck (or save it)
3. Effort - saving energy for the next task or the long shift ahead
4. Pleasurable – I don't mean fun. I mean maybe it's because the task is tedious in keep repairing the issue or they are in autopilot, as things are not challenging any more.

We had found a solution to a problem that was causing issues for a year. Honestly, it was wrong! I ask you what problems are you having that are neat, plausible, and mainly wrong to get the job done?

Now is the time to stop and reassess the problems, issues and challenges that are causing you a headache. If you can't see the solution, be brave and ask for someone with fresh eyes to look at it. As they say, two heads are better than one.

And I never did find out when temporary becomes permanent…

It's like trying to found the answer to the meaning of life or the Ultimate Question of Life, the Universe, and Everything? Actually, the last question has been answered, it's 42, but when temporary becomes permanent is the ultimate question to life, the universe and everything.

Path

Shortcut, Loophole - or Skewed perspective?

To me a shortcut is something that goes wrong with the loophole. A loophole is a method to save either time and/or effort.

A skewed perspective is where one person thinks that they are doing a great job in saving time or effect or money. From another view, the person would see that in the end it will cost just as much time, money and/or effort.

There is a story by Arthur C. Clark about loopholes. The summary of the story goes something like this:

It begins in the form of correspondence between the President of Mars and the Secretary of the Martian Council of Scientists, regarding the discovery of atomic power (in the form of atomic bombs) by humans.

They are concerned that once humanity's current war is over (they have been monitoring Earth's broadcasts), humans will use atomic power and rockets to breach interplanetary space and pose a threat to Mars.

A remote monitoring station is set up by Mars on the Moon to monitor Earth's progress. Finally, they send a fleet of 19 battleships along with a warning to Earth that one city will be destroyed every time a rocket leaves Earth's atmosphere.

Earth agrees to stop experimenting with rockets when they realize their broadcasts are being intercepted. Ten years pass without any further rocket experimentation, while the Martians plan for the extermination of the human race, believing that Earth will always be a threat to them.

The next letter, beginning "Mars is a mess!" is sent from Mars by a human named Lieutenant Commander Henry Forbes, and reports to Earth upon the destruction of Martian civilisation due to a nuclear attack from Earth.

The previous letters had been recovered from the ruins of the capital. Rather than experimenting with rockets, humans had perfected matter transmission and beamed their bombs directly over the Martian cities. Forbes is hopeful that rocket experiments will resume soon, as he finds being "beamed" across space to be uncomfortable.

People will shave effect and time to get the task and/or jobs done, it is a matter of taking everybody on the right path.

Lucky?

While people bend the rules, and get away with it, they will keep shaving and shaving until they get bitten or simply can't shave anymore.

I love stories and while travelling there is nothing better than to listen to these to pass the time. One of my favourite stories is War of the Worlds.

> *Ogilvy, the astronomer, assured me we were in no danger. He was convinced there could be no living thing on that remote, forbidding planet. "The chances of anything coming from Mars are a million to one," he said. "The chances of anything coming from Mars are a million to one - but still they come!*

I'm sure you could think of some silly answers like divorce the partner or go on a wild binge and spend it all.

I received compensation for my accident. With two ex-wives, there isn't any left. I enjoyed some of it, as it allowed me to live in a beautiful little hamlet of Grove, east of Canterbury, Kent on the Stodmarshes. These marshes were used by the monks of Canterbury Cathedral to look after horses and breed them (so the 'Stod' is actually stud).

The day the settlement was reached was like playing 'Deal or No Deal'. I sat in one office and the insurance company in another. At about 10am a piece of paper was handed to my solicitor (enough to buy a luxury car and a nice holiday). Without me saying anything, it was rejected.

Over the next six hours, figures and answers where batted to and fro. The solicitor informed me that she didn't think the insurance company would move anymore after 6 hours of offers made.

I was asked what I wanted to do. I could take the money or reject it. Before I even gave the answer, the solicitor went through what would happen if I rejected the offer and it went to court.

The sum offered would then either decrease back to the original offer or increase at least double to what was being offered. I'm not much of a gambler (I like a few pence on a Saturday, during the football season, but that is it). I decided to take what was on offer.

The solicitor disappeared to tell the insurance company. Before I knew it a man came in shook my hand and said 'well done young man, you're now very rich! Goodbye'. I just sat there like a guppy at feeding time.

Am I rich? I will let you answer that question.

Sometimes, during presentations, the subject of money comes up and I tell the story about the day I played 'Deal or No Deal'. I then follow it up by asking the question about the richest man in Babylon (inspiration from the book by George Samuel Clason's book The Richest Man in Babylon).

Who is the richest man in Babylon? The king sitting in his tower with all the wealth, but no health or the beggar in the gutter with no wealth, but with all his health?

Money is nice, but it's no good if you haven't got your health!

A Goldfish Called Bob!

There is the African proverb it goes a little like this: What are fish the last to recognise?

The answer is in fact **water** as they are in it all the time. A bit like yourself at work - you get used to your environment.

It is not until someone picks you out of your environment that you see the good and bad habits that you've picked up over time. It needs to be highlighted that it is not just about the bad, but the good too.

The trouble is, good habits are not recognised or rewarded enough, because they are either taken for granted or just expected.

Things are normally only understood from one point of view and it blinds us from other perspectives because nobody else is in our water.

There is a story concerning Alexander the Great and a yogi, and how they see the world.

Alexander the Great was busy conquering the known world when he saw, on the banks of the Indus River in today's Pakistan, a naked guy sitting in the Lotus position contemplating. "Gymnosophists" (gumnos = naked, sophistes = philosopher) the Greeks called these men.

"What are you doing?" asked Alexander.

"Experiencing nothingness," answered the yogi. "What are you doing?"

"Conquering the world," said Alexander.

Then both men laughed, each thinking that the other must be a fool.

"Why is he conquering the world?" thought the yogi. "It's pointless."

"Why is he sitting around doing nothing?" thought Alexander. "What a waste of a life."

The questions need to be asked who had a pointless life and who had a waste of life? The answer depends on the pond you are swimming in, and the information that is let in.

Cultures and Standards

I stay in many hotels around the United Kingdom for work and one thing I see advertised is breakfast and how good it is. Apart from Fish and Chips or Roast Beef Dinners the Brits are known to love a good fry-up. It's part of our culture!

Now when I stay in England this is known as a? .... Full English

When I stay in Scotland this is known as a? .... Full Scottish

When I stay in North Ireland this is known as a? .... Ulster Fry (In the Republic of Ireland it's known as a full Irish).

When I stay in Wales this is known as a? .... Cooked Breakfast

Are we not one country called the United Kingdom? So, shouldn't it just have one name?

The fried breakfast is a bit like your work as each depot or department. All have their own culture, but actually they are the same company.

A bit like me at my place of work. Different shifts, departments, mills – one company, but different cultures.

These cultures at work drive the standards on site.

The New Zealand rugby union team are the current world champions and some say they are the greatest team in history of the sport.

In my place of work (Recycled Fibre Plant) we were known as

the elite of the papermakers because nobody in the paper industry could match the quantity or quality we were producing. Not only were we the best, we were the safest as well, because we hadn't had a lost time incident in five years.

We needed to have someone to shake us out of the blindness of what was happening on site; to show us the bad and the good.

To highlight this point, I'm going to take you back to the 2001-2002 football season the winners were my team - Arsenal - by seven points. Sir Ferguson of Manchester United moaned that they were the better team, even though Arsenal had beaten them 3-1 at home and 1-0 at Old Trafford. To Sir Ferguson's comment Arsene Wenger (as I write this North Londoners greatest manager for another two years) said 'Everyone thinks they have the ***prettiest*** wife at home.'

So, do you have the prettiest partner at home?

To bring this quote into the workplace, people think they are the best. The people they work with are good, but not as good as them. The tools and machinery are OK, but overall, they do the job and we get the job done.

Now consider this.

You get to the airport with family in tow for your annual holiday.

After waiting around you board the plane, find your seats and as you sit down you see this gentlemen out of your window.

What would most people do? They would nudge each other, take photos, have a giggle and post on social media for the world to see and comment.

You order your drinks ready to fly. The engineer finishes and the plane gets to the runway and takes off. During the flight as the plane is in the air you look out the window and says 'it's still attached!' and then takes another sip of holiday juice, as the other half replies 'I can see that!'

Once the plane lands and everybody disembarks the man in the group turns round and says in the most macho fashion, 'I knew there was nothing to worry about!'

Now bring this sort of action into the workplace. You see someone working in a potentially unsafe manner. What do most people do? They walk on by. Even if they do enquire, they get an answer like 'go and take the longest walk off the shortest of piers!'... piers!'... Or words to that effect. So, we don't have conversations. As, Cloughie once said 'it only takes a second to score a goal'! It only takes a second to save a life and I'm not on about 'OI'! I'm on about 'Come on, let's have another look before we proceed' That is the difference between going home or not.

The picture was taken from the Daily Mail newspaper website. The story is that contrary to popular belief, it is not duct tape the worker was using. It is believed the item is actually speed tape, widely used in aviation to carry out quick repairs on flights so as to avoid delays.

Speed tape is an aluminised pressure-sensitive tape used to do minor repairs on aircraft and racing cars. It is used as a temporary repair material until a more permanent repair can be carried out.

It has an appearance similar to duct tape, for which it is sometimes mistaken, but its adhesive is capable of sticking on an airplane fuselage or wing at high speeds, hence the name.

The point is, the passengers were uneasy about the presence of the speed tape but nobody actually said anything – even though for all they knew it could've brought down the aircraft. Speaking up is vital!

So was I compliant at work? Yes. I wore my PPE like everyone else, wore personal locks, ready to lock off any equipment if we had a breakdown and followed the rules. Well, 99% of the time I followed the rules, as rules are black and white. None of us break rules, but we do bend them.

Machinery is key, because most companies are about production (if your industry is not machinery focused then it's about delivery). While people rush around trying to fix issues on sites, you potentially see the worse of human nature, as you see people point fingers and scream and shout at each other. No matter what the product is, it must be delivered to the customer on time and within budget.

Even though there were many departments within the company, the main focus was on PM6. A monster of a machine, it was six metres across and about 100 metres long. It produced about a tonne of paper a minute. It is an all-consuming beast, but while it's running 24/7 it is a happy place. If it is not running it is hell, as people rush around to sort out the issue.

Because for us at the mill, while the sheet is off the company is losing money and potentially customers.

Blinkered

In the 1970s there was a children's programme called the Magic Roundabout (if you're old enough you can sing the tune and remember the characters).

So, what has the Magic Roundabout got to do with health and safety at work? The Magic Roundabout is a metaphor for bits of machinery or trouble spots around the workplace that are always causing issues.

Our Magic Roundabout was that machine from Sweden. It only ran every two or three weeks as production dictated. We knew though that during the shift when running the return system, it would block.

The 25 of us across the shifts developed the system. One man in the switchgear room and two men at the bottom by the hatches. I have been asked many times if I remember the first time I put my arm into the abyss. I can honestly tell you I can't remember.

I was watching a program about Piper Alpha (the oil rig that blew up in the North Sea in 1988, 120 miles north east of Aberdeen killing 167 people). The program makers interviewed Geoff Boland who was in the control room during the disaster. During his interview, he said:

'I didn't go to work thinking I was particularly unsafe. It was an environment you soon got used to'.

Like Geoff I felt safe at work. Having worked in the paper industry, you get used to the environment (mainly hot, humid and noisy). How many of us get used to the environment and the things happening around us?

We become blinkered to any issues, because the more the issue is present, the more we become blind to it.

As the Magic Roundabout turns, it spits out bullets and hopefully at the moment you have managed to dodge them – but how much longer can you dodge them for? A day, a week, a month or a year. Hopefully it will be a life time.

Moment in Time

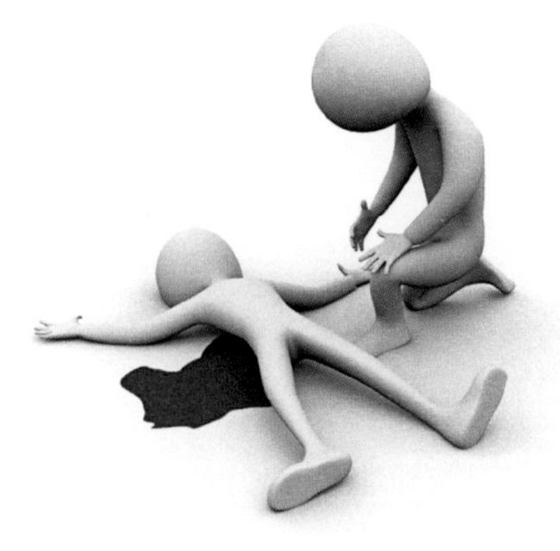

When I look back to that time, I remember laying on my back with the medical team rushing around in the assessment room before being taken to theatre. It reminds me of the story of the sparrow by The Venerable Bede.

> *"The present life of man upon earth, O King, seems to me in comparison with that time which is unknown to us like the swift flight of a sparrow through mead-hall where you sit at supper in winter, with your Ealdormen and thanes, while the fire blazes in the midst and the hall is warmed, but the wintry storms of rain or snow are raging abroad.*
>
> *The sparrow, flying in at one door and immediately out at another, whilst he is within, is safe from the wintry tempest, but after a short space of fair weather, he immediately*

vanishes out of your sight, passing from winter to winter again. So, this life of man appears for a little while, but of what is to follow or what went before we know nothing at all. If, therefore, this new doctrine tells us something more certain, it seems justly to be followed in our kingdom."

Do we really appreciate moments in time? Do we really appreciate being in the moment at all?

Technology

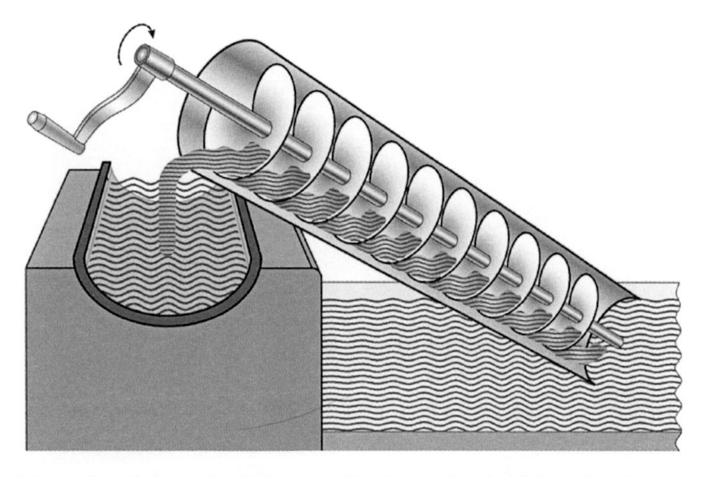

At this point, it is only right to talk about the Archimedes Screw (or screw conveyor). Some may even call it an auger.

The history of the Archimedes screw (screw conveyor) actually dates back before Archimedes by some 350 years (around 625 BC)! Some scholars believe it was used in Assyria (Assyria would today cover Northern parts of Iraq, Syria and Iran, and the Southeastern region of Turkey).

Since my accident I have tried to see if I could find any more machines like these. To be honest, I haven't found any. What I have found is a set-up where the feed starts at the top and gravity is used to feed downwards to other screws.

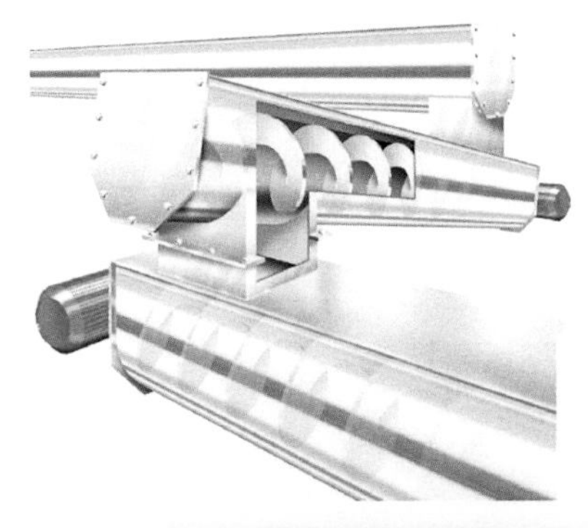

Fig 6 Crossover Screw Conveyor

Last year, I did finally find a set-up like the one I used - but what a change! The hatches have been removed and to gain access you have to open the end of the crossover point. How technology moves on, develops and evolves over time.

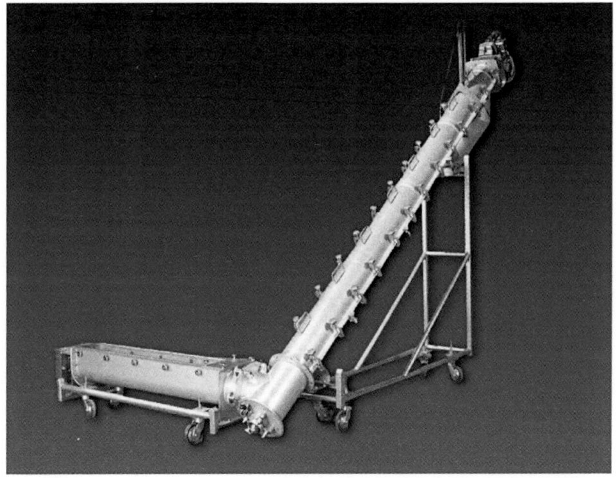

Fig 7: Modern day crossover screw conveyor

Perfect Storm

So, why did my accident happen? It was a convergence of circumstances like many accidents or incidents. The perfect storm.

- **Miscommunication**. We regularly had issues with radios failing. Working in buildings with steel frames mean radio signals are lost. Radio engineers came in to fit better amplifiers, but the radios still had issues. You take it on the chin that during your shift you are going to have to repeat yourself several times.
- **Lack of procedures**. We never had any procedures for the Archimedes Screw. We were very good at planned shuts as we had permits to work. On the run, we were not so good.
- **Lack of leadership**. This was prevalent in all layers of the organisation. At times, when you are having a bad day, you just nursed the machinery through until the next shift came on. You simply hand over the issue to the next person. This situation does the rounds. Going up the food chain of the organisation there was two issues. One was a certain manager who use to tap his watch and whistle at you like a dog. The other was the shift log. Everyone in the control room was expected to fill in what had happened over the shift. This was then read by management to understand what had happened over the days, weeks and months. When the HSE came in they picked up that in the six months before

my accident, the plant had 35 blockages and nothing was done to address the issue, as the machine suppliers passed the buck between themselves. The machine was brought originally as management believed it would never block.

- **Over confidence**. In other words, too much trust in the people you work with. I had worked with the guys for five years and when you work with people every day you get to know what they are good at (or not). Everybody works to their strengths. These guys become family as you see and work with them so much. My colleague John D was a Royal Engineer (previously), he was the plant's safety rep – he wouldn't get it wrong, would he? He did, unintentionally of course, and I have no ill will against him.
- **Complacency.** Because you get use to the things including machinery and people around you. You lose the fear with the machinery as you work on them every day and develop with the people a sixth sense (understanding of working).

Complacency

Complacency cannot be seen, but you can feel it.

I want to take you back to your first day at work. Do you remember that feeling in your chest like butterflies? Once settled, they soon disappear. As you get used to the job, you also get used to the people, the overall culture and what is expected of you.

Once the last of the butterflies disappear, a fuse is lit. For the majority of people at work the fuse just keeps on burning and these people retire and live a long and uneventful life.

The others (approx. 1%, if you calculate from the HSE stats) see the fuse burn out and explode in their face, marking them for the rest of their lives, to reflect and learn from that incident.

I don't know how to get the butterflies back, but to keep them there is vital if you can, as they keep you on your toes and aware of your surrounds, and potentially get you home every day.

If you are looking for some inspiration watch the James Bond film The Spy Who Loved Me. During the film James Bond is trying to track someone down called Fekkesh, and as ever he is in the arms of a beautiful woman.

James Bond: *Where is Fekkesh?*

Felicca: *You are very suspicious, Mr Bond*

James Bond:Oh, I find I live much longer that way

If you can have that little bit of suspicion in everything you do, it could be the difference between going home that day or not.

Complicit

We were actually complicit in the situation at work, as we didn't speak up enough to management about the issues we were having.

Yes, I did have a conversation (before it was installed) about the new machine. I said it was not right regarding the screw conveyors as to me it made more sense to have flatbed conveyors. These were like the ones used over on PM6 to return broke (waste paper).

I was politely told that the machine was going to be installed, and we were going to have to get on with it.

So, the 25 of us across the five shifts got on with it.

All this time after my accident I do wonder if all the manufacturers actually knew that there were issues reported elsewhere concerning their bits of kits.

Come to think of it, how many times are conversations about situations at work, kept amongst a team of people? Issues discussed and points raised which never make it to the ears of management. These conversations are complicit actions, as people know something is wrong but they do nothing about it, just like in the speed tape scenario. They fear losing their job or that no-one will listen, or worse that they might be ridiculed. They carry out these unsafe actions, as they feel powerless to do anything.

Getting that open culture and encouraging people to speak up would make the world of different to any workplace, not just in

Health and Safety terms but in team morale and management awareness.

Do you remember the story of the Emperor's New Clothes from your childhood? As it is about being complicit and being part of the hyper-normalisation that go on in organisations where everybody knows the truth that the systems are failing, but everybody plays along to keep the myth going and keep the status quo intact, in fear of the unknown. In the story, the pompous king is persuaded by mischievous tailors that a 'magnificent' and extremely expensive suit they have produced for him can only be seen by clever people. In fact, there is no suit at all, so when the king wears the suit, the king is actually naked. Everyone can see that he's naked, but nobody speaks out for fear of breaking the status quo and looking stupid. Nobody wants to be the first to question the claim.

So, an entire population is persuaded to adopt a completely false belief - based on exploiting people's individual pride, fear of embarrassment, and reluctance to be a lone voice of disagreement.

The tale ends with the intervention of a small boy, who, unaware of the widely publicised mythical claims of the tailors, loudly pronounces the king to be naked, so exposing the sham.

Experts (All of Us)

Captain Hindsight (from South Park) and his sidekicks Shoulda, Coulda and Woulda depicts the inspiration to talk and write about hindsight and how we see it.

Hindsight is a beautiful thing as it makes us experts after the event and that is the trouble it is always after the event. The more beautiful thing is foresight and having the courage to speak up and spot the opportunities around us.

My own hindsight is that I should have walked up the two flights of stairs and 100 metres down the machine floor to put a lock on in the switchgear room. I did the same as everybody else, and I paid the price for my actions.

Whilst in hospital, my shift mates visited me. As we chatted, you could see in their eyes that look of thank Christ it was you and not us!

When I look back, I realise it actually *was* them more than me! I never really saw anything, apart from when I pulled the last remaining bit of arm out of the screw. I witnessed nothing more than that. My mates saw far worse, especially Darren who for that hour and a half knelt over me and stopped me from bleeding to death.

Of course, we do learn from the should ofs, could ofs and would ofs - but foresight and forward thinking is far more useful. And the biggest skill for all workers (including managers and directors) is

LISTENING to both sidesof the situation and not being judge and jury. It's easy to make assumptions. Listening allows us to learn.

Think about the bishop, priest and the ladle (now this is funny...)

A bishop invited a young priest to dinner. During the meal, the priest noticed some signs of intimacy between the bishop and his housekeeper.

As the priest was leaving, the bishop said to him quietly, "I can guess what you are thinking, but really our relationship is strictly proper."

A few days later the housekeeper remarked to the bishop that a valuable antique solid silver soup ladle was missing - since the young priest's visit - and so she wondered if he might have taken it. "I doubt it, but I will ask him," said the bishop.

So the bishop wrote to the priest: "Dear Father, I am not saying that 'you did' take a solid silver ladle from my house, and I am not saying that 'you did not' take a solid silver ladle from my house, but the fact is that the ladle has been missing since your visit..."

Duly, the bishop received the young priest's reply, which read: "Your Excellency, I'm not saying that 'you do' sleep with your housekeeper, and I'm not saying that 'you do not' sleep with your housekeeper, but the fact is that if you were sleeping in your own bed, you would by now have found the ladle."

Genius!

Foresight is also about spotting opportunities around us. Sometimes they are not obvious, but they're there right in front of you.

A story about opportunity:

A fellow was stuck on his rooftop in a flood. He was praying to God for help.

Soon a man in a rowboat came by and the fellow shouted to the man on the roof, "Jump in, I can save you. The stranded fellow shouted back, "No, it's OK, I'm praying to God and he is going to save me." So, the rowboat went on.

Then a motorboat came by. The fellow in the motorboat shouted, "Jump in, I can save you." To this the stranded man said, "No thanks, I'm praying to God and he is going to save me. I have faith." So, the motorboat went on.

Then a helicopter came by and the pilot shouted down, "Grab this rope and I will lift you to safety." To this the stranded man again replied, "No thanks, I'm praying to God and he is going to save me. I have faith." So, the helicopter reluctantly flew away.

Soon the water rose above the rooftop and the man drowned. He went to Heaven. He finally got his chance to discuss this whole situation with God, at which point he exclaimed, "I had faith in you but you didn't save me, you let me drown. I don't understand why!"

To this God replied, "I sent you a rowboat and a motorboat and a helicopter, what more did you expect?"

Watch this video and see how good you are at spotting opportunities.

How well did you do?

It's the little things which people take no notice of. These things are dismissed as they seem unimportant.

The trouble is all that happens is these little things are swept under the carpet. That is, until you can't fit anymore under the carpet and it collapses like an avalanche down a mountain.

These little things can be dealt with at handovers or team meetings if they are indeed essential to a business.

We had handovers. Unfortunately some were "War and Peace" and some were "The Very Hungry Caterpillar". It depends on what is seen as important (or not).

It comes back to how you see the world.

> *'Things are always unnoticed until they have been noticed…' (Sir Richard Broadbent)*

He was actually talking about the financial issue with Tesco back in 2014. You can take this quote and apply it to accidents/ incidents at work. How many things are missed on site until thump! Something is damaged and/or someone is injured.

All of a sudden everything that was under people's noses as was part of the furniture is now painfully obvious;

How did we miss that?

Why did we miss that?

When did we miss that?

Even Harry Houdini made mistakes. His great jail escape was simple: Houdini kept his regular street clothes on, and after being thoroughly search was locked inside the cell with no audience. Within a specified time limit he would free himself. Every time he accepted this challenge he escaped.

But one time his luck almost ran out…

> *A small town in Britain invited Houdini to accept a challenge to escape from their jail. This jail cell was the pride of their police force. It was a brand-new cell constructed with the most recent technology available. Houdini accepted the invitation and was locked up.*

After the room was cleared Houdini took off his jacket and got to work. Hidden in his belt has a flat 10-inch metal tool Houdini had designed to pick locks. He began to work on the lock, but soon he began to realise there was something different about this lock that he had not seen before.

For thirty minutes, he worked and worked and got nowhere. He tried everything he could think of to open the lock with no luck. His confidence began to fade and his thoughts shifted to the possibility of defeat.

Time continued to pass. He continued to struggle against the lock. Covered in sweat and passing the point of exhaustion he collapsed against the door. To his surprise the door gently swung open, revealing that it had never been locked in the first place...

Workers

Fig 8: Warehouse

Fig 9: Dock

Is this good working, as they have saved the companies they work for money by using the resources available?

At your place of work what department or shift or depot it is?

It is interesting when showing these types of pictures to delegates in organisations the feedback that comes forth about their work colleagues. I was in the West Midlands on a night shift and asked what shift this was? The answer came: “That’s us!”…

My good lady Cheryl has a terrible phobia about spiders. She will jump at fluff on the carpet that is 20 feet away. I have begged her to see someone about it, as she is 5ft 4 and spiders are only an inch (25mm) at most. It is totally irrational to me!

The other day we had the same conversation, when a spider run across the living room. After screaming and me stop laughing, as I pointed out she was irrational she piped up, do you remember the 25th November 2000? I replied I think so. She continued "That day a plonker went to work in a paper mill, and while they had a blockage and to clear it he stuck his arm in a hatch of a screw conveyor without it being isolated properly and then needed a 16 hour operation to reattach it! That is bloody irrational".

Ok, she had made her point.

To one it is rational, but to another it's irrational.

Fig 10: Warehouse again *Fig 11: Dock again*

So, are these employees working rational or irrationally? To them they are working rationally, but to others it's irrational. People who can't see the dangers are stuck in the goldfish bowl and need a little shake outside to show them their error.

Shared Memories

Teams work a lot on shared memories and experience to get a job done. In psychology, it is called "transactive memory" and it was first developed by Daniel Wegner.

Weger gave his hypothesis in 1985 as a response to earlier theories of "group mind" such as group thinking. He described a transactive memory system as being a mechanism through which groups collectively encode, store, and retrieve knowledge. Transactive memory was initially studied in couples and families where individuals had close relationships, but was later extended to teams, larger groups.

According to Wegner, a transactive memory system consists of the knowledge stored in each individual's memory combined with metamemory, containing information regarding the different teammate's domains of expertise. Just as an individual's metamemory allows him to be aware of what information is available for retrieval, so does the transactive memory system provide teammates with information regarding the knowledge they have access to within the group. Group members learn who knowledge experts are and how to access expertise through communicative processes. In this way, a transactive memory system can provide the group members with more and better knowledge than any individual could access on his own. How many times do people call upon you to answer a question at work and you think 'they should know that already?!' as they've done the job years. How many times does this actually happen to you?!

That spiral of asking and answering is transactive memory; the team acts as a kind of 'library of knowledge'.

So, how do teams fail in tasks that require transactive memory? It is a combination of things that include the non-sharing of information. One of the group fails or refuses to share knowledge they know will help get job done or gets the group up to speed, because the person with the knowledge is insecure of their position.

The team just doesn't have the full knowledge between them to complete the task. This could be just a newly installed machinery. Yes, the team has the theory of how to get the machinery fixed, but the team needs the practical knowledge too.

It could be that momentary slip in knowledge - that on the tip of your tongue moment. You know the answer, but it just doesn't come, and the team is stuck until either the answer comes or they make it up just to get the job done.

To quote Ken Blanchard: "None of us is as smart as all of us"

Picture a cage containing five monkeys.

Inside the cage, hang a banana on a string and place a set of stairs under it.

Before long, a monkey will go to the stairs and start to climb towards the banana.

As soon as he touches the stairs, spray all of the monkeys with cold water.

After a while, another monkey makes an attempt with the same result - all the monkeys are sprayed with cold water.

Pretty soon, when another monkey tries to climb the stairs, the other monkeys will try to prevent it.

Now, turn off the cold water.

Remove one monkey from the cage and replace it with a new one.

The new monkey sees the banana and wants to climb the stairs.

To his surprise and horror, all of the other monkeys attack him.

After another attempt and attack, he knows that if he tries to climb the stairs, he will be assaulted.

Next, remove another of the original five monkeys and replace it with a new one.

The newcomer goes to the stairs and is attacked.

The previous newcomer takes part in the punishment with enthusiasm.

Again, replace a third original monkey with a new one.

The new one makes it to the stairs and is attacked as well.

Two of the four monkeys that beat him have no idea why they were not permitted to climb the stairs, or why they are participating in the beating of the newest monkey.

After replacing the fourth and fifth original monkeys, all the monkeys that have been sprayed with cold water have been replaced.

Nevertheless, no monkey ever again approaches the stairs.

Why not?

Because as far as they know that's the way it's always been around here…

Habits

2500 years ago, a 23 year old by the name of Alexander the Great was conquering the known world.

On his travels around Persia, Afghanistan and the northern part of India he was quoted as saying 'Remember upon the conduct of ***each*** depends the fate of ***all.***'

This quote is as relevant today as it was back then, because it is what is put it place that will have an effect on everybody in an organisation.

My accident was eighteen months in the making from the time the order was placed for the machine to the time it bit me. Any one of us could of stopped, taken a step back and had discussions to stop what we were doing.

I'm not just about talking about the bad things and habits, its talking and recognising the good things happening and habits.

It is the conscientious people at work are killed or maimed because they want to be seen as loyal, reliable employees. They go beyond the boundaries to get the job done or kept on track.

Talking to delegates I say that they don't even have to stop a job unless it is needed - just slow down, take a step back and take your time. The more you slow down the more you see – and it's less likely there'll be an accident or surprises. I joke to delegates the more they slow down they can drink more coffee (or tea) and dunk their favourite biscuit (it is normally a chocolate hob nob). It also helps health and safety be easier to digest on toolbox talks and courses.

If you need a hint to help get people buy in to this idea, you can use the quote by Douglas Adams: *"If it looks like a duck, and quacks like a duck, we have at least to consider the possibility that we have a small aquatic bird of the family anatidae on our hands."*

Or…

If it **looks** unsafe, **sounds** unsafe and **feels** unsafe. It's unsafe stop the job!

Fail Safe

I have talked about hindsight and foresight, which is person limited. One thing that is needed are fail safes. The systems that stop incidents and accidents happening.

In the workplace, they are called risk assessments, safe systems of work and policies, that kind of thing. However, these are not always the safety net people think they are.

The 2017 Oscars are the case in point. It enshrined everything of the old adage "if it can go wrong it will go wrong."

- The handing out of the wrong envelope.
- An actor caught in the headlights regarding what to do next
- An actress losing patience, as she was kept out of the loop.
- The same actress reading what she *thought* was on the card – not what actually was
- Total panic backstage as they realized the wrong film was read out

At each stage, there was a failsafe in place, but every one of them failed, causing one of the biggest talking points of 2017.

Once fail safes fail people jump to find a scapegoat because someone must carry the can. For the Oscars, it was the accountants who handed out the envelopes. Actually, do we really learn in identifying a scapegoat and piling all the blame on them?

Back to the F1! When Williams mixed up the tyres on Bottas' car Rob Smedley said 'What we shouldn't do is blame individuals. We need to go away as a group of people and look at what happened'.

I was asked once who I blame for my accident. I replied that in a way we were all to blame for what had happened because we all had a responsibility to stopping what had happened, but we failed in that task, whether employee, manager or director and suppliers.

The gentleman who asked the question was puzzled by the answer, because to him someone had to carry the can surely?

Once the safety measures fail the hunt for the scapegoat must begin and unfortunately, emotions get in the way of really understanding what happened.

Scapegoats become stereotypes, because even if there hasn't been an incident or an accident, people have labels placed upon them.

Take the workplace for example. There are lots of stereotypes to be found, especially around working behaviours when things look a bit 'off'. How many times have you seen pictures of company employees posted on social media with some comment about the person in the photo being called an idiot etc?

Most of the time they are not idiots. Actually, I'm going to argue that they are clever as they are using the resources they have at hand to get the job done! To ensure the customer is happy, their boss is happy and the money keeps rolling in.

I proved it earlier with the forklift trucks picture and the quayside photograph. Here is another picture to prove the point.

Fig 12: Many hands make light work

Safe behaviour in the workplace will only come about if the tools and resources are in place. There also needs to be consistency. Once a strategy is in place, it has got to stay in place. I have seen too many CEOs or Directors change things because they didn't like the previous CEO's or Director's ideas.

Let me illustrate my point:

Once upon a time there lived a man named Diogenes who was as wise as any man who ever lived. People admired Diogenes for his intelligence, but he was a rebel who lived his whole life in opposition to others. In winter, he walked barefoot in the snow. In summer, he rolled in hot sand. When people asked him why he did such things, he explained that he was hardening himself against discomfort.

One day Diogenes was walking across the agora, the ancient gathering place, heading toward the Acropolis. But he walked backward, and as men ambled past him, they laughed and whispered and wondered why he would do such a thing.

Diogenes pretended not to notice the whispers until a large crowd had gathered around him. Then he turned and said, "You laugh at me because I am walking backward? But you lead your entire lives backward and cannot change your ways so easily as I."

With that, he turned on his heels and began to walk in a normal fashion.

Another day, Diogenes walked past a stream, where he noticed a young boy scooping out water and drinking from his hands. He stopped to admire the child and sighed to know the child had outdone him in simplicity. He tossed away his cup and he never used it again, for Diogenes believed that no one ought to have more things than he truly needed. He believed no one needed much.

In fact, instead of living in a house, Diogenes slept in a tub made from a barrel. He also rolled from place to place in that barrel. His greatest fear was that he would awaken one day to discover he lived in a palace while all those around him lived in barrels.

Diogenes spent most days lying in that tub, with his face in the sun, waiting for anyone who wished to stop by and ask him a question. People came from far and wide seeking his wisdom.

But one day, after he had been dispensing wisdom for a while, he climbed out of the tub and began to walk through the streets. It was a bright, sunny day, but Diogenes carried a lighted lantern and peered around as if he were searching for something.

As people passed, they wondered what he could have lost, but most were wary of asking. At last, a young man who did not know of the great man's reputation innocently asked, "Sir, why would you carry a lantern in this bright light?"

Diogenes looked at the young man and said, "I'm searching."

"Searching for what?"

"An honest man," Diogenes answered. "Alas, I don't see a single one in sight."

Word of Diogenes' wisdom spread to the king, Alexander the Great, who had been tutored by Aristotle when he was very young. When Alexander succeeded his father, Philip II of Macedon, to the throne, he was a learned man. By the time he was 30, Alexander had created one of the largest empires in the world, stretching from the Ionian Sea to the Himalayas. He was known throughout the world as one of the great military leaders. But he believed Diogenes might have something to teach him, and so he decided he would travel to Corinth to see the philosopher.

Naturally, the important men of Corinth came out to see their king and to offer him praise and gifts and compliments. All the important men came, that is, except Diogenes, who did not bother to leave his tub.

Since Diogenes would not travel to see the king, the king decided he must go to see the wise man, and his men told him he would likely find the man in his barrel tub.

"He's a most peculiar man," they warned him.

Alexander did not care. He was looking for wisdom, and he understood that wisdom sometimes comes in strange packages.

It happened that day that Diogenes was beside his tub, lying in a pool of sunshine on the grass, enjoying the warmth and light after a long winter. When he heard the sound of the crowds moving toward him and the blare of trumpets announcing the arrival of the king, he looked up.

He squinted at the sight of the procession coming toward him.

When Alexander spotted the great philosopher, he leaped from his horse and ran forward, so excited was he to hear Diogenes' words.

He quickly reached the man and smiled down and said, "Diogenes, I have heard many stories of your wisdom and I've come to ask if there is anything I can do for you. Please, feel free to ask me for anything."

Diogenes looked up, thought for one moment, and said, "Yes, yes there is."

"Anything," the king answered, pleased to do what he could.

"Stand a little to the side, please," Diogenes requested. "You're blocking my sun."

When they heard these bold words, Alexander's guards gasped, but the king was so surprised by the answer that he could only laugh. After that, he admired Diogenes even more. He stepped aside, and Diogenes thanked him, sincerely.

"It is a wise man who keeps his word," Diogenes said, and as Alexander and his men turned to ride home, the king smiled.

He turned to his servant and said, "I can tell you this, if I were not Alexander, I would like to be Diogenes."

If CEOs and directors don't keep their word, the worker at the sharp end loses hope. In the end, they just do what they need to do to hit their targets and go home with a wage to support their family.

Common Sense

Before you carry on with this chapter why not complete a common-sense test at:

http://brainfall.com/quizzes/do-you-have-common-sense/#HJZNTsnNZ

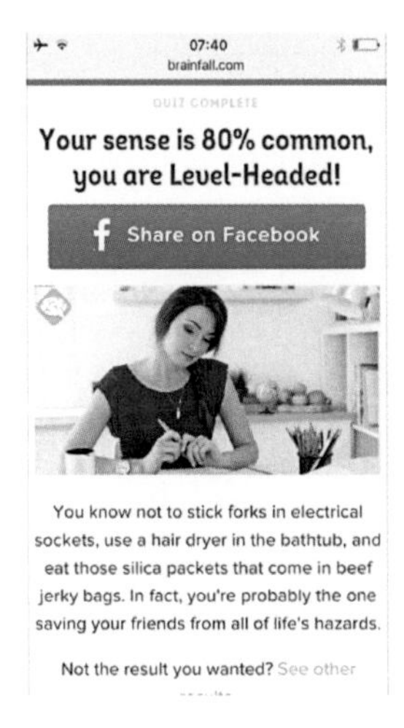

This wouldn't be much of a health and safety book without a chapter on common sense.

But what is common sense? If you were to do a search in the Encyclopaedia Britannica, common sense according to the 18th and

19th century philosophers Reid, Ferguson, Stewart (and others), it is the perception of the average unsophisticated man.

If you were to search Wikipedia, its opening line states the following: Common sense is a basic ability to perceive, understand, and judge things that are shared by ("common to") nearly all people and can reasonably be expected of nearly all people without need for debate. Basically, common knowledge or belief amongst a group of people.

Reading further, Aristotle narrowed it down to five specialised sense perceptions: sight, hearing, taste, touch and smell. Most people use all of these, every moment of the day, unless they have been perhaps ill or injured.

Illness and injury are not the only things that slow the senses. Tiredness plays its part too, whether not sleeping at home (between shifts) or resting at work properly (not taking your break).

The lack of having full use of our senses plays a part in bringing us down. Look at how many sportsmen or women that have admitted that they failed because they were not 100% ready or fit to compete due to recovering from a virus.

Employees are not Olympians, but they are your best asset and it is great to see many organisations now roll out wellbeing programs for everyone working for them.

There is a little story from Japan about common sense, involving a priest and a huntsman:

Once there lived upon the mountain called Atagoyama, near Kyoto, a certain learned priest who devoted all his time to meditation and the study of the sacred books. The little temple in which he lived was far from any village; and he could not, in such a solitude, have obtained without help the common necessaries of life.

But several devout country people regularly contributed to his maintenance, bringing him supplies of vegetables and of rice. Amongst these good folk there was a hunter, who sometimes visited the mountain in search of game.

One day, when this hunter had brought a bag of rice to the temple, the priest said to him "Friend, I must tell you that wonderful things have happened here since the last time I saw you. I do not know why such things should have happened in my unworthy presence. But you are aware that I have been meditating, and reciting the sutras daily, for many years; and it is possible that what has happened to me is due to the merit obtained through these religious exercises. I am not sure of this. But I am sure that Fugen Bosatsu comes nightly to this temple, riding upon his elephant. Stay here with me this night, my friend; then you will be able to see and to worship the Buddha."

"To witness so holy a vision," the hunter replied "would be a privilege indeed! Most gladly I shall stay, and worship with you."

So the hunter remained at the temple. But while the priest was engaged in his religious exercises, the hunter began to think about the promised miracle, and to doubt whether such a thing could be. And the more he thought, the more he doubted.

There was a little boy in the temple, an acolyte, and the hunter found an opportunity to question the boy. "The priest told me," said the hunter, "that Fugen Bosatsu comes to this temple every night. Have you also seen him?

"Six times, already," the acolyte replied, "I have seen and reverently worshipped Fugen Bosatsu." This declaration only served to increase the hunter's suspicions, though he

did not in the least doubt the truthfulness of the boy. He reflected, however, that he would probably be able to see whatever the boy had seen; and he waited with eagerness for the hour of the promised vision.

Shortly before midnight the priest announced that it was time to prepare for the coming of Fugen Bosatsu. The doors of the little temple were thrown open and the priest knelt down at the threshold, with his face to the east. The acolyte knelt at his left hand, and the hunter respectfully placed himself behind the priest. It was the night of the twentieth of the ninth month, a dreary, dark, and very windy night; and the three waited a long time for the coming of Fugen Bosatsu.

But at last a point of white light appeared, like a star, in the direction of the east. The light approached quickly, growing larger and larger as it came, and illuminating the slope of the mountain. Presently, the light took shape, the shape of a divine being, riding upon a snow-white elephant with six tusks. And, in another moment, the elephant with its shining rider arrived before the temple, and there stood towering, like a mountain of moonlight.

Then the priest and the boy began to repeat the holy invocation to Fugen Bosatsu. But suddenly the hunter rose up behind them, bow in hand; and, bending his bow to the full, he sent a long arrow whizzing straight at the luminous Buddha, into whose breast it sank up to the very feathers.

Immediately, with a sound like a thunder-clap, the white light vanished, and the vision disappeared. Before the temple there was nothing but windy darkness. "O miserable man!" cried out the priest, with tears of shame and despair, "O most wretched and wicked man! What have you done? What have you done? "

But the hunter received the reproaches of the priest without any sign of compunction or of anger. Then he said, very gently: "Reverend Sir, please try to calm yourself, and listen to me. You thought that you were able to see Fagan Bosatsu by reciting the sutras. But if that had been the case, the Buddha would have appeared only to you, not to me, nor even to the boy. I am an ignorant hunter, and my occupation is to kill; and the taking of life is hateful to the Buddhas. How then should I be able to see Fugen Bosatsu? I have been taught that the Buddhas are everywhere about us, and that we remain unable to see them because of our ignorance and our imperfections. You being a learned priest of pure life might indeed acquire such enlightenment as would enable you to see the Buddhas; but how should a man who kills animals for his livelihood find the power to see the divine? Both I and this little boy could see all that you saw. And let me now assure you, reverend sir, that what you saw was not Fugen Bosatsu, but a goblin intended to deceive you perhaps even to destroy you. I beg that you will try to control your feelings until daybreak. Then I will prove to you the truth of what I have said."

At sunrise the hunter and the priest examined the spot where the vision had been standing, and they discovered a thin trail of blood. And after having followed this trail to a hollow some hundred paces away, they came upon the body of a great badger, transfixed by the hunter's arrow. The priest, although a learned and pious person, had easily been deceived by a badger.

But the hunter, an ignorant and irreligious man, was gifted with strong common sense; and by common sense alone he was able at once to detect and to destroy a dangerous illusion.

Common sense boils down to what is common to our senses. What I did on the 25th November 2000 may have not been common sense to some, but it may have been common sense to others.

To me, what is not common sense is when vans and lorries break the speed limit with their company logo plastered all over them. When this happens it just makes you think are these companies really concerned with health and safety, or do the employees driving these vehicles believe in the message being delivered?

Bias

There is a theory called "Theory U". I learned about it while on a facilitation course for LEGO® Serious Play, (it's a great method to break down the barriers of normal meetings or courses as it is about 100% participation from everybody, as everybody has an opinion worth sharing). The theory was developed by Otto Scharmer.

It is about opening minds, hearts and will. For me it's about understanding how you see yourself, how you see others and how you see the world. Once you conquer these layers you can move on to others you live and work with and understand how they operate with themselves, others and the world around them.

Bias. It's a good thing. It generates debate amongst teams and it allows growth, as ideas are grown and nurtured. We do have to be careful that the wrong ideas are not allowed to take root though, but they should be aired, to ensure they are corrected.

If you want to understand biases, look out the window (it's my bias that you have a window, as you read this book by the way). Walk around for 10 minutes (again, my bias that you can walk) or sit and watch (my bias being to assume you can actually see the world around you). As you do this, label the people moving around you. That is your biases showing themselves.

Another way to see biases is to ask people what football team they support, who they normally vote for and how they voted in the Brexit referendum.

My answers are:

I support Arsenal (due to that great 1979 cup final and Alan Sunderland scoring the winner), vote for whoever puts across the best vision for Britain and Brexit (It will be a good thing in the end, as we have been founded as a nation of traders with the world, not just the neighbours (Europe). Whilst we established an empire [now the commonwealth], Europe was trying to conquer each other.

As you read my answers, it would have been great to see your face, and seen the smiles and frowns, that went across it. What judgements – biases – you instantly developed.

I'll finish with a neat little story that show bias in a less obvious way:

During a visit to NASA in the 1960s, President John F. Kennedy (or maybe Johnson or Nixon. It depends on the version you heard or read first) noticed a janitor moping the floor. He interrupted his tour, walked over to the man and said, "Hi, I'm the President of the United States. What are you doing?"

"Well, Mr. President," the janitor responded, "I'm helping put a man on the moon."

Right Signage!

As I was writing this book something landed in my lap that I just couldn't leave out. It was the picture of Mike Pence (Vice President of the US) touching a bit of NASA equipment and the sign right in front of him saying

'DO NOT TOUCH'

When I read the article, it stated that it was actually okay to touch, as it needed a clean anyway before the next process in producing the Orion rocket. Yeah, right! If anyone else (employee or visitor) had touched it, would they be working at or visiting NASA today? Probably, not.

Having the right signage (and the right amount) on sites is paramount, not just for employees, but visitors too. On my travels, I have seen the extremes of both. Too much signage and the message gets lost. Too few signs and you're left guessing.

The question needs to always be asked, is the sign relevant? I always found the signage placed on the screw conveyors after my accident ironic. I don't need to worry about it now, as the machine was decommissioned in 2008; the new owners classed it as obsolete to requirements.

Having the right wording is important too:

DON'T THINK OF YOUR FIRST KISS

I bet you remembered that first kiss or even a kiss! Now think about the signage at work when it says DON'T on the sign. The word DON'T might as well not be there, as the subconscious totally ignores the word.

Another issue is having it in the right place, so it is there in your face without you missing it. Big supermarkets spend lots of money on observing where people look as they shop for products.

If I remember rightly from my Sainsbury's days, the average shopper will look at the middle of the aisle as they move around the supermarket. Once they are moving down the aisle they will only look at eye level to waist. A field of about 2foot (60cm) or so.

The next time you go shopping have a look at where certain brands are placed. Safety signage can learn a lot from supermarket psychology about shoppers.

Communication

"There are known knowns. These are things we know that we know. There are known unknowns. That is to say, there are things that we know we don't know. But there are also unknown unknowns. There are things we don't know we don't know." - Donald Rumsfeld

When Rumsfeld first said this, I can remember most of the comedians at the time taking the mick out of what was said. There were even grammar experts debating whether the statement was right or wrong.

Whilst researching this book, I came across an argument in Wikipedia that actually Rumsfeld had missed an element, as Psychoanalytic philosopher Slavoj Žižek pointed out. He states the fourth element was, the unknown known, which is where we deliberately refuse to acknowledge that we know what is happening around us.

It is interesting how people who can't explain things very well are dismissed in any discussion, even if their point is valid. I got an 'F' in my English GCSEs. I hated English, as a subject I found it boring. Looking back, I think it was because it didn't float my boat, it wasn't exciting! Nearly 30 years on I'm sitting here writing a book. Oh, the irony! Now history was another thing, if I could have spent my days at school studying History I would have been like a pig in the proverbial, because I could debate the subject, unlike English.

History was a subject I loved and still do. You can get a lot out of learning from the past; you can see the future approaching, as History always repeats itself, like a long-playing record.

It's easy to dismiss people in what they are talking about. Whether it's because they are boring, you think they are beneath you or there are more pressing things to be done.

Research by Merriam states that how we communicate is the following:

Body language is 55%

Tone is 38%

Words is 7%

How we communicate is important; how we listen and observe to these signals are just as important, if not more so. We are throwing out information all the time and only the shrewd amongst us really do understand what communication is.

Falling Trees

In the Solomon Islands in the south Pacific some villagers practice a unique form of logging. If a tree is too large to be felled with an axe, the natives cut it down by yelling at it. (I can't lay my hands on the article, but I swear I read it...) Woodsmen with special powers creep up on a tree just at dawn and suddenly scream at it at the top of their lungs. They continue this for thirty days. The tree dies and falls over. The theory is that the shouting kills the spirit of the tree. According to the villagers, it always works.

Ah, those poor naive innocents. Such quaintly charming habits of the jungle. Screaming at trees, indeed. How primitive. Too bad they don't have the advantages of modern technology and the scientific mind.

Me? I yell at my family. And yell at the telephone and the lawn mower. And yell at the TV and the newspaper and my children. I've been known to shake my fist and yell at the sky at times.

My next door neighbour yells at his car a lot. And this summer I heard him yell at a stepladder for most of an afternoon. We modern, urban, educated folks yell at traffic and umpires and bills and banks and machines - especially machines. Machines and relatives get most of the yelling.

I don't know what good it does. Machines and things just sit there. Even kicking doesn't always help. As for people, well, the Solomon Islanders may have a point. Yelling at living things does tend to kill the spirit in them. Sticks and stones may break our bones, but words will break our hearts...

While at times we might scream, shout and yell at other people whether close or not, plus objects big or small and it makes us feel better about the world as it is now off our chests and the troubles lifted off them shoulders. Have we really made them issues disappeared? Probably, not! All we have done is passed the problem to somebody else to solve or carry.

Openness

How do we get everyone on side to achieve that safer culture, but more importantly, that *open* culture? The story Steve Jobs (Apple) told about teamwork will help. If you haven't heard or read it, please find it below:

I've always felt that a team of people doing something they really believe in reminds me of something from my childhood. When I was a young kid there was a widowed man that lived up the street. He was in his eighties. He was a little scary looking. And I got to know him a little bit. I think he may have paid me to mow his lawn or something.

One day he said to me, "Come on into my garage I want to show you something" and he pulled out this dusty old rock tumbler. It was a motor and a coffee can and a little band between them. And he said, "come on with me." We went out into the back and we got just some rocks. Some regular old ugly rocks. And we put them in the can with a little bit of liquid and little bit of grit powder, and we closed the can up and he turned this motor on and he said, "come back tomorrow."

And this can was making a racket as the stones went around. And I came back the next day, and we opened the can. And we saw these amazingly beautiful polished rocks. The same common stones that had gone in, through rubbing against each other like this (clapping his hands), creating a little bit of friction, creating a little bit of noise, had come out these beautiful polished rocks.

The more we rub against each other and not sit in our silos (departments) the more we became the polished item.

I have seen too many managers use the divide and conquer theory with teams, so in the end teams fight amongst each other and the managers blame their staff for the failure of the team.

Teams at times do need a kick up the arse to achieve more, because they sit on their laurels - they think there is no more to achieve.

Teams know when they are working for a poor leader. It was defined when I was watching about the Shah of Iran's party in the desert in the 1970s and one of the commentators said you can tell a weak person (leader) because they are only worried about status and money and then they go blind!

Leadership (I'm going to call it "Guidance")

Leadership means thinking outside the box. It could it be what any Health, Safety, Behaviour or Cultural change needs to be put in place to win and be a success. That crazy idea that seems mad, but in the end it was obvious.

I ran an under 8s football team for Woodcoombe in Sittigbourne (I had played for them over the years) during the 2002/03 season. I had about eighteen boys of various skills and experience.

We were to play in the Eastern Division of the Medway youth league. My personal goal for the season was to have a respectable season and finish in the top half. If there was a chance to win the league they needed to get a minimum of four points out of the two games played each week.

The boys were split into two teams. I knew from my past that other managers would usually keep all the best players in one team and the weaker ones in the other.

If I was to do the same the points for each game would be shared; the best playing teams would probably draw most of the games, with the odd win or loss along the way.

The second game would be unpredictable. As a team, I felt it best that the strongest defenders would play in the first game and draw the game because we knew it would be against the harder team.

The second game had the best strikers, as we knew the second game would have the weaker players in the opposite team. I was lucky as both teams had good midfielders so, these were split evenly between the two sides. The plan worked we won the league and were finalists in the consolation cup.

Why have I told you about how I won a youth football league? Because we won by thinking outside the box. We knew the issue ahead (win the league), we knew the possible hurdles (better teams) and we applied a solution that could win.

One other secret? I told the boys that even though they played in separate games, they all played for Woodcoombe Under-8s, so they all won as a team, drew as a team and lost as a team. There was a collective amongst them, and me, as we got to know each other. We knew the strengths, weaknesses, opportunities and threats.

How many organisations are split up into small chunks (silos)? And these chunks are all treated differently? Organisations have a tendency to isolate departments, teams, people, from the rest, to highlight best practise or worse preforming.

There is never that togetherness, safe together, healthy together and/or change together. Leadership is about creating a team of people that win, draw and lose together and they trust each other. Elections are won in the centre ground - but that means shared perceptions of competence and charisma, hopes and fears, not a nauseating mixture of Marmite and marmalade. As stated once by Mark Mardell on The World This Weekend, commenting on the Labour Leadership battle between Corbyn, Burnham, Cooper and Kenall

From hindsight, I know we get in a rut where life is comfortable. Life is flowing down the river without a care in the world and any objects blocking the way are just pushed to one side.

The second part that I picked out was about how elections are won. Actually, any leader of any level in an organisation has to get a balance of these four elements:

Competence: being seen to do the job or know what they are talking about.

Charisma: being able to charm the group into believing that the job can be done.

And finally, the last two are together as the leader must show that they understand the hopes and fear of the group, and have empathy.

Look at Brexit and the General Election (2017). Neither side really won because not enough of the people believed one way or the other that Brexit was the right thing to do. The General Election of 2017 showed no party inspired the right message, as they only appealed to their own natural supporters.

It's about the understanding that we can't please all the people all of the time, but we can please most of them.

Leadership is about great marketing of people and getting them to where they want to be. Great management is about control over the systems of the organisation and ensuring they are reliable and robust for the task.

There is a story about how the wolf pack hunt and protect each other, and it goes something like this:

There is a wolf pack of 25 wolves: the first three are the old or sick, they give the pace to the entire pack. If it was the other way round, they would be left behind, losing contact with the pack. In case of an ambush they would be sacrificed. Then come five strong ones, the front line. In the centre are the rest of the pack members, then the five strongest follow them.

The last one is alone - the Alpha. He controls everything from the rear. In that position he can see everything and decide the direction. He sees all of the pack. The pack moves according to the elder's pace and help each other, watch each other."

Change

It's easy to resist change. Leaders have to be the ones that cracks the nut to Nirvana and find the prefect culture of health and safety at work.

The Government has even joined in with its own Nudge Unit, nudging people in the 'right' direction. Like the sign I came across on the way home on the A299 where it said, 'Please take your rubbish home, others do!'

My take on the prefect culture is as simple as keeping the organisation's essence and integrity intact. The trouble with most change projects that fail is that they are imposed too quickly.

These change projects fail to win the hearts and minds of the troops on the ground, and with the loss of traction at the bottom it then feeds up the chain of command, until it collapses.

The change has got to be needed as well, and not just done on a whim, because someone thinks it will be a good idea to have a change. Take the recent decision to make Doctor Who a woman! It has split the Whovians (Doctor Who fans) in half. It makes no sense to some as they see the Doctor as male because he is a Time Lord not a Time Person or a Time Lady. It didn't help to win these fans back, when one of the writers said on Nicky Campbell's phone in on BBC Radio 5 Live the Monday after the decision that they chose a woman because it was time.

Comments like this do not help keep the essence or integrity of the program for those fans. It destroyed the years of devotion that they have spent following a beloved interest and fighting for it to return to the screens following years off air, and then someone hijacks it and puts a nuclear bomb under it.

Any change needs to be started with a single step, not a giant stride. As Loa Tzu, the ancient Chinese philosopher and writer said, 'The journey of a thousand miles begins with one step'. The change needs to be slowly evolving, mopping up as it goes. It soaks up the hearts and minds of those involved.

The changes also need to be seen to be taking effect, so easy wins can be observed taking place and have the backing as desired. Take the butterfly or moth (or any other insect) that undergo complete metamorphosis. These insects have four life stages: egg, larva, pupa and adult. Each stage is different, but each stage also has a change for its use.

The Butterfly Story

A man found the cocoon of a butterfly. One day, a small opening appeared. He sat and watched the butterfly for several hours as it struggled to force its body through the little hole. Then it seemed to stop making any progress. It appeared stuck.

The man decided to help the butterfly and with a pair of scissors he cut open the cocoon. The butterfly then emerged easily. Something was strange. The butterfly had a swollen body and shrivelled wings. The man watched the butterfly expecting it to take on its correct proportions. But nothing changed.

The butterfly stayed the same. It was never able to fly. In his kindness and haste the man did not realise that the butterfly's struggle to get through the small opening of the cocoon is nature's way of forcing fluid from the body of the butterfly into its wings so that it would be ready for flight.

Like the sapling which grows strong from being buffeted by the wind, we all need to struggle sometimes to make us strong.

When we teach others, it is helpful to recognise when people need to do things for themselves.

Change is good, but it can't be rushed and it must have a purpose to help achieve a greater good. Any change is awkward to start off with (try writing in your other hand). But once you keep applying the new change, it finally becomes automatic.

Legacy

In his book "Legacy", James Kerr talked about the New Zealand rugby union team leaving the shirt in a better place.

It's a shame the England teams following that great team of 1966 have not followed on by leaving the shirt in a better place…

That will change in 2018 as the England team go to Russia, to win the World Cup and overturn 52 years of hurt, with three lions on their shirt.

We ourselves don't talk about legacy, what we want to leave our families and friends, or how we want to be remembered at work.

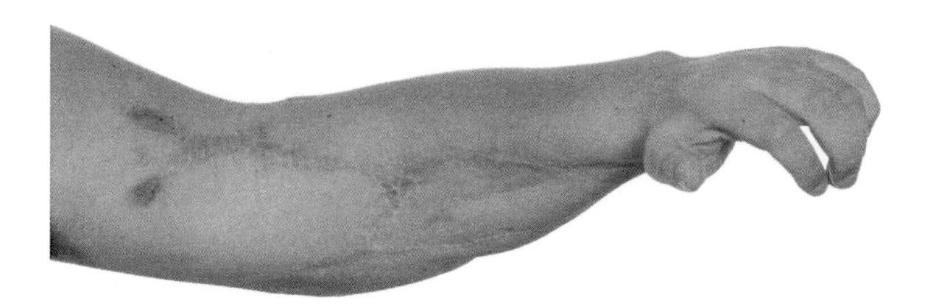

This is my beautifully scarred arm. It runs from my armpit to my palm. It's the first thing I see in the morning and the last thing I see at night. All because I thought I was doing the right thing!

Monica Lewinsky talking at a TED conference said about her affair with Bill Clinton "Not a day goes by that I am not reminded of my mistake, and I regret that mistake deeply."

As I said, yes, I am reminded of my mistake every day. How the hell can you run from a scar the length of your arm! But do I regret it? No, I don't because, why should I? I have appeared on the BBC's 999 program and the local TV news on BBC & ITV. I was part of a BBC Kent radio show at the St. John Ambulance Kent Headquarters, concerning the importance of first aid at work and home.

I was in the printed press locally and nationally, and talked all over the world about the days leading up to the day, the day itself and the time after afterwards. Plus, now it includes the lead up to that day too. I couldn't tell you the number of people who have seen or heard the story, but if it has stopped just one person and their family and friends going through what I did, then it has been worth it.

I feel hurt that I have put people I love through the ringer at times, but I hope they can see that I have turned a horrific situation into something positive.

Accidents and incidents will happen and as long as they are not major we have to accept it. It is how the human race has got to where it is by learning from mistakes.

When the winds blow

Years ago a farmer owned land along the Atlantic seacoast. He constantly advertised for hired hands. Most people were reluctant to work on farms along the Atlantic. They dreaded the awful storms that raged across the ocean, wreaking havoc on the buildings and crops. As the farmer interviewed applicants for the job, he received a steady stream of refusals.

Finally, a short, thin man, well past middle age, approached the farmer. "Are you a good farmhand?" the farmer asked him. "Well, I can sleep when the wind blows," answered the man.

Although puzzled by this answer, the farmer, desperate for help, hired him. The little man worked well around the farm, busy from dawn to dusk, and the farmer felt satisfied with the man's work. Then one night the wind howled loudly in from offshore. Jumping out of bed, the farmer grabbed a lantern and rushed next door to the hired hand's sleeping quarters. He shook the little man and yelled, "Get up! A storm is coming! Tie things down before they blow away!"

The little man rolled over in bed and said firmly, "No sir. I told you, I can sleep when the wind blows." Enraged by the response, the farmer was tempted to fire him on the spot. Instead, he hurried outside to prepare for the storm. To his amazement, he discovered that all of the haystacks had been covered with tarpaulins. The cows were in the barn, the chickens were in the coops, and the doors were barred. The shutters were tightly secured. Everything was tied down. Nothing could blow away.

The farmer then understood what his hired hand meant, so he returned to his bed to also sleep while the wind blew.

Dining Room Table

Life - what's it all about? Let me tell you, but you need to be sitting down. Concentrate on these next couple of pages, as I explain what life is.

The Thursday morning after my accident I finally sat upright in a chair. I was watching the world go by in Canadian wing (this is where the badly burned pilots in World War II were treated and become the famous Guinea Pig Club at the Queen Victoria Hospital at East Grinstead), when a lady sat down and introduced herself as Dr. Alexander, the trauma counsellor for the hospital.

She asked if I could talk her through what happened on Saturday. "Not really, I know what happened and why I'm here", I replied.

She asked again about the events of Saturday. Like you do, I took a deep breath and got ready to tell her the events as quick as I could. But, all I did was take the breath, because Dr. Alexander then dropped a bombshell by saying "Please can you get a picture of a dining room table in your head?"

Pardon?! My face must've been a picture!

You do what she says, because you are in no mood to resist, let alone fight. So, I got a dining room table in my mind. With the table in my head, Dr. Alexander asked about my family. I explained about Nick (my wife at the time) and Georgia, my daughter. Then I talked about Mum and my little sister Faye. I told her that my dad had died when I was 16.

She asked if I could now turn my family into objects and put them onto the table. I can't remember what the objects were, but they were things that wouldn't look out of place on a dining room table.

We moved on to friends and we repeated the process. Work and then hobbies and interests followed, we must have talked for 45 minutes or so. I was asked what the table looked like, I said to me it looked like a Christmas table full and overflowing.

She then asked me to pick up everything on the table, and everything just floated in the air. With everything picked up off the table, she commanded that I smash everything on the floor. Everything hit the floor and smashed into a million pieces. With that I cried, as I realised that split-second decision had nearly cost everything.

As I wiped away the tears, Dr. Alexander said that today we are going to pick those pieces up. I must admit the pieces started to come together again slowly. All these years on, every now and again I find the smallest of pieces that need to go back on the table somewhere.

That table doesn't look like it used to, but it's my table!

The question is - what could you really afford to smash from your dining room table?

"Life can only be understood backwards; but it must be lived forwards". - Soren Kierkegaard

As we get older this is how we understand life. We reflect on the successes and the errors that have happened during our journey of life. As we look back, we continually move on with our lives, correcting and learning from our actions, whatever they are.

Over the last five years I've thought deeply about what happened that day and before it. The best way to clarify what happened is the following story:

A farmer and his son had a beloved stallion who helped the family earn a living. One day, the horse ran away and their neighbours exclaimed, "Your horse ran away, what terrible luck!" The farmer replied, "Maybe so, maybe not. We'll see."

A few days later, the horse returned home, leading a few wild mares back to the farm as well. The neighbours shouted out, "Your horse has returned, and brought several horses home with him. What great luck!" The farmer replied, "Maybe so, maybe not. We'll see."

Later that week, the farmer's son was trying to break one of the mares and she threw him to the ground, breaking his leg. The villagers cried, "Your son broke his leg, what terrible luck!" The farmer replied, "Maybe so, maybe not. We'll see."

A few weeks later, soldiers from the national army marched through town, recruiting all the able-bodied boys for the army. They did not take the farmer's son, still recovering from his injury. Friends shouted, "Your boy is spared, what tremendous luck!" To which the farmer replied, "Maybe so, maybe not. We'll see."

Little Things

It's the little things in life that have the biggest impacts on our lives.

If you have ever stayed in hospital, you will know that your meals are selected one day in advance. So, the first day on Canadian wing (Thursday) I hadn't selected my meals as I had been in Davies Ward, the high dependency ward.

Lunch went without a hitch as it was sandwiches, but tea (supper, dinner whatever you want to call it) was a little more difficult.

As tea was placed in front of everyone, and you could see them lifting the lid to glimpse what they were having. As I hadn't picked my meal, I was the last to get tea. It was placed in front of me and I was hoping for Macaroni Cheese, or something soft and easy to eat. But as I lifted the lid, there it was - pizza and green beans! I hate green beans, but I do like pizza (not those horrible frozen six inch things though). I gave the nurse that puppy eyed look of you got anything else? She stared back with that mother look of "tough"!

I looked at the fork, knife and spoon and gave the nurse that puppy eyed look again of could you cut it up for me? And she returned that look again.

So I stared back at the fork, knife and spoon and though how the hell am I going to eat this pizza?! I considered just picking it up and eating it with my hand, but that was out of the question because if the cheese dropped on to my new dressing from yesterday I would be in big trouble.

Back to the fork, knife and spoon. The spoon was a no go, so it was easy to dismiss it outright. The knife wouldn't work as I knew as soon as I started to cut it, it would fly across the room like a frisbee.

All that was left was the fork. Maybe I could just cut it up with the pressure of its side? Something as easy as eating pizza became a mountain to climb, but I did it.

It really is the little things that we take for granted which have the biggest effect on us. One of the biggest things in life is that little scribble at the bottom of letters or that order form. Yes, I'm talking about your signature.

It took me about three weeks to prefect it, writing it over and over again on notebooks with my good hand.

Here is my challenge to you, before you read on. Please have a go at writing your signature with the opposite hand to one you usually use.

....

....

....

....

How did you do? This is another challenge for you. See if you can beat this gentleman?...

As I say it's the little things in life that have the biggest impacts.

Here is part of the journal of one of the Society of Estate Clerks of Works of Winchester, England in 1908.

He was the best machinist in the district, and it was for that reason that the manager had overlooked his private delinquencies. But at last even his patience was exhausted. He was told to go, and another man reigned in his stead at the end of the room.

And then the machine, as though in protest, refused to budge an inch. It just wouldn't work! Everyone who knew the difference between a machine and a turnip tried his hand at the inert mass of iron. But the machine, metaphorically speaking, laughed at them. The manager sent for the discharged employee who left the comfort of the "Bull" parlour and came.

He looked at the machine for some moments, and talked to it as a man talks to a horse. He then climbed into its vitals and called for a hammer. There was the sound of "tap-tap-tap," and in a moment the wheels were spinning, and the man was returning to the "Bull" parlour.

*And in the course of time the mill-owner had a bill:–"To mending machine, £10. 10s." The owner of the works, being a poor man, sent a polite note to the man, in which he asked him if he thought tapping a machine with a hammer is worth ten guineas. And then he had another bill:—**"To tapping machine with hammer, 10s.; to knowing where to tap it, £10; total, £10. 10s."***

And the man was reinstated in his position. So grateful was he that he turned teetotal and lived to a great and virtuous old age. The moral is that a little knowledge is worth a deal of labour! To gain that knowledge we have to put the effort into things to gain that knowledge to succeed.

The Ripple

Let me now go through the four groups that are hit straight away by a major incident or accident. On average, the minimum number of people affected is 44! This is the bare minimum that is affected in a major incident or accident.

Family:

They are hit by the tsunami of the event. Mum, Faye and Nick saw me on the Sunday afterwards (I didn't see them as I was out for the count). Mum saw me and turned to both Faye, Nick and the nurse and exclaimed that I was her baby! Even though I'm the eldest and I was 27.

In my second week, Nick helped me to have bath. I had to keep my arm out of the water, so washing was not easy. She had to wash my hair three times as it was full of blood from Saturday and where I laid.

Friends:

Some friends will stay and see you through. Some will cross

the road out of embarrassment and others act like rabbits caught in headlights!

These friends are great, as they can't quite 'get' what has happened. One friend (I won't embarrass them, but I have known him since infant school) use to talk to my arm, never to my face. I let it go, as it is always good to see people. One day I said to him that I was up here, pointing to my face. He simply paused and said "I know" before looking back to my arm.

Work:

The workplace is interesting as the great morale that has been on site disappears as quickly as clicking your fingers.

People on site start to question and point fingers, not just at others, but at themselves too. How could this have happened? We didn't we see it coming?

Whilst questions are flying, a great bunch of people from the HSE (Health and Safety Executive) come on site and put people on the rack (mentally) like medieval torture, because they need answers.

For us it was the wrong piece of machinery, none of us were truly trained on it, the machinery suppliers were not taking responsibility for the blockages and it had become 'the done thing' to clear the blockages that way. The last shift to run the machinery at 3am a fortnight beforehand had two guys up to their armpits in the hatches clearing blockages.

EMS (Emergency Medical Services):

The final group and more often than not, the forgotten group. They come on to sites put people back together and do what they need to again somewhere else, until they go home.

I even had the police raid Asda (about four miles away) to get the ice for my arm. I understand the policeman had to knock customers out of the way to get the ice to preserve my arm.

Sixteen-and-a-half years on, that ripple keeps on going touching people I have never met or ever will do.

"A woman is like a tea bag - you can't tell how strong she is until you put her in hot water".

Eleanor Roosevelt said that as first lady of the United States in the 1930s. Men are too. We were all dipped that day, it's just some were stronger than others. Everybody did their job in saving me and my arm.

As I lay on the floor I was asked who my next of kin are. I said Nick, Mum and my sister, but I insisted that they needed to phone Mum, as I knew Nick would panic - she was looking after Georgia at the time too. Mum got the call. Like all mums she is the rock of the family. She said to give her ten minutes to walk to my house and she would break the news to Nick.

How would your family react to the call or knock on the door?

How would you cope with being the one who has to make the call or knock?

Coping

Mental illness is a big thing in the Western World. For me it's because the Western World thinks it has to move at 1000mph and it if you don't achieve that you're a failure.

I have always coped the best I can with whatever life throws at me. I think it's because I'm a Londoner from birth. I was brought up with stories of the war, especially about the dog fights above Nan's and when the V1 bomber took out several streets at the bottom of hers. It was real boys stuff listening to this, especially imagining the Spitfires and Messerschmitt BF 109s doing their deadly dance in the sky.

I would say we were brought up to just have a go and see if you can do it, and if you can't then ask for help. So, I would say I'm a stubborn, independent old basket and I guess it has got me in trouble a few times too.

When Dr. Alexander had finished with me, she said she would

"see me the following week". I must admit I went back to bed that afternoon and had nightmares about what I done. These lasted for a week and in the end, they become pleasant (or should I say I turned them into fond memories to which I remember two of them to this very day).

The two I remember to this day are sticking different parts of my body into that hatch to see what happened. The other was being a scrap piece of paper and going through the whole process in the Recycled Fibre Plant, the paper machine and finally conversion and becoming a whole new piece of A4. Weird how the brain copes with things…

Seeing Dr. Alexander the next week she was stunned by how well I had progressed. What I had done was learn to dance with the demons in my mind.

Laughing is the other thing that has got me through, either watching comedy or listening to it. There is nothing more satisfying than having a bloody good laugh to ease the stress of life.

Torie, my second wife, told me that in a roundabout way. She told me once "the trouble with you is you put a bloody good act on in front of people, but they don't see you at home!"

She was right. But as my Nan Ciss said to mum while I was in hospital "You'll never know what that boy went through!". She's right. As much as I try and explain what happened, I can't get over to you all the smells, sounds, sights and feelings I experienced that day, so you could never really understand.

I don't see myself as any different from anybody else, apart from having been involved in a major accident and having ground breaking surgery. The only thing I do differently is build time into anything I do, because doing things one handed takes time!

So, how do you cope? I would say be yourself and accept life has taken a new direction. Most importantly, laugh and smile as the world will come with you and support you. As the Black Knight from Monty Python and The Holy Grail said, "Tis but a scratch!"

Home – Who's Waiting for You!

This is Cheryl, my good lady at home. Before we go off to work every morning, we always have a kiss and a cuddle, tell each other we love each other. When we get home we repeat the process and talk about our days at work, as happens every day around the world. It hurts me to think that on that day I went to work at 6am whilst Nick and Georgia were asleep.

They were expecting me home that evening at about 7.30. Nick knew I would walk in with a takeaway (like a lot of families up and down the country Saturday night is takeaway night). We would have a quick kiss and cuddle before Nick would dish up tea and I would go and find littl'un, who would more than likely watching Tellytubbies on the sofa.

So, who is expecting you home?

I hope this book is inspiring you and others that you work with to work that little bit more safely in your workplace. The people you love really are the ones who pick up the pieces!

Hopefully this story will explain it better:

Once upon a time, there was an old man who used to go to the ocean to do his writing. He had a habit of walking on the beach every morning before he began his work. Early one morning, he was walking along the shore after a big storm had passed and found the vast beach littered with starfish as far as the eye could see, stretching in both directions.

Off in the distance, the old man noticed a small boy approaching. As the boy walked, he paused every so often and as he grew closer, the man could see that he was occasionally bending down to pick up an object and throw it into the sea. The boy came closer still and the man called out, "Good morning! May I ask what it is that you are doing?"

The young boy paused, looked up, and replied "Throwing starfish into the ocean. The tide has washed them up onto the beach and they can't return to the sea by themselves," the youth replied. "When the sun gets high, they will die, unless I throw them back into the water."

The old man replied, "But there must be tens of thousands of starfish on this beach. I'm afraid you won't really be able to make much of a difference." The boy bent down, picked up yet another starfish and threw it as far as he could into the ocean. Then he turned, smiled and said, "It made a difference to that one!"

Adapted from The Star Thrower, by Loren Eiseley (1907 – 1977)

Time

If you think you can't change or inspire others, please read about Frances Oldham Kelsey and her battle to keep Thalidomide off the streets of America, while it ravaged the rest of the world.

If you are not from the health and safety profession, and need more convincing that now is the time to change and work more safely! Please see the numbers below and the toll it is having on people and their families. Below are the stats from the HSE for 2015/2016 with my interpretation attached to the figures below. As each and every number is someone hurt and has had suffered pain at work, and who has affected their family and friends.

Those killed at work 144 – this is an airliner (like easyJet or Ryanair use) once a year flying off with a full plane of people and never returning home.

RIDDOR (Reporting of Injuries, Diseases and Dangerous Occurrences Regulations) where an accident is reported to the authorities 72,702 – nearly the full capacity for a Manchester United home game once a season being injured at work and a quarter of this crowd will have life changing injuries.

Cost to the country every year for injuries and new cases of illness £14.1 billion – every man, woman and child (based on a population of 65million) paying £216.92 every year to help ease the pain and suffering.

30,400,000 days lost due to injury and illness – That is nearly every person in full time employment having 12 hours off this year or a day and half depending on your hours worked.

I will be honest and admit I have been a plonker, I should of spoke up more and had them conversations. Luckily, I have lived to tell the tale.

I let time slip through my fingers and I paid the price. Please use your time wisely, and go home every day to your love ones in one piece. As the last thing I want to read is that someone else has suffered at work and another family is picking up the pieces.

Lines of Enquiry

I attended the Confederation of Paper Industry (CPI) conference in 2015. Hastam (Health & Safety experts) was there and they ran a session about accident investigation. My accident was used as the basis for it.

I took away the questions that were asked during the two sessions. It is only right that I address each one as best I can.

Machine – Purchasing:

The machine was purchased because the Recycled Fibre Plant could not keep up with the paper machine when on 75% or more recycled pulp. To extend the production run by up to 12 hours the machine was installed. According to the manufacturers in Sweden, it was guaranteed to not block up…

Change of pulp – wet and/or dry:

We used bales of pulp made for our sister mill in Sittingbourne or one of the M-Real mills in France. As these bales could be weeks old, there was no telling of the moisture content of each bale, as depending on where it was in the stack would be subject to whether it was wet or not. Sometimes you could have a dry bale in the middle of a stack because of space and/or runs.

PUWER assessed (Provision and Use of Work Equipment Regulations 1998):

To be honest I cannot remember if it was assessed, as this would have been carried out by the safety team on site.

Modifications to the machine:

There were no modifications carried out to the machine until after the accident. Afterwards, an interlocking key system was installed, maglocks were fitted to the hatches and mandatory signs stuck to the screw casings.

How operators were trained:

I cannot remember being trained by anybody in depth. If we were it was around the control panel to de-wire the bales. Unblocking the screws were by trial and error. We had no safe systems of work.

Reporting structure – why didn't management know?

In the control room was an up-to-date shift log and it was the responsibility of the person in the control room to write any issues down. These shift logs were used for both morning and monthly meetings. Part of the HSE investigation looked at the shift logs and they went back 6 months to find around thirty-five blockages reported.

Management of radios:

Radios were kept in the control room on charge and there was another base in the Wetlap hut and over in the warehouse. We had issues with signals and an engineer came in to fit a better amplifier, but we still had problems hearing. So in the end, it just became part of the working day - sometimes you had to repeat yourself a lot...

Operator involvement in machine selection, design, safe system of work and when last reviewed:

On a personal note, my opinion was never sought when it came to safe working. I had mentioned that I didn't think that that particular machine was the right machine for the task at hand. It made more sense to use flat-bed conveyors to return the pulp. As far as I know the review was only carried out once my accident happened.

Safety concerns raised about blockages:

I personally didn't raise any and neither did anybody else. I did think to myself though that the screw conveyors were the wrong set up, as flatbed conveyors would have made more sense in returning pulp to a re-pulper.

Common practice to remove blockages:

Yes, it was common practice to unblock screw conveyors like we had done. The process was one person seeing to the isolation in the switchgear room and everyone else doing the unblocking with air lines, iron bars and hands.

To help remove the second blockage we did run the screws backwards, and to my mind this was the first time it was done.

Management awareness around cleaning:

Yes, they were aware of the way we unblocked the screws because at times some even helped with the housekeeping. As far as I was concerned we were never pulled up about the way we were clearing the blockages.

System to monitor water content in bales:

No, there was no monitoring of water content in bales.

How were the hatches used after blockages?

The hatches were closed with metal wedges to keep them shut.

Maintenance:

Maintenance was carried out on the macerator teeth about three or four months before my accident.

Permits to work the system and for use in unblocking:

Yes, we had a permit to work system, and this was used comprehensively for major shuts on site, as well as breakdowns to motors, pumps and wire changes (mats that dewater pulp).

Why was the lock-off so far away?

Sorry, I can't answer that question. All I can think is it was a suggestion from the manufacturers, as this would have been the closest switchgear room to the machine.

How were other blockages cleared?

Depending on what machine and where the blockage was, different methods were used to unblock it.

How did company culture influence operator behaviour?

The company was very much about ensuring machines run at top speed, and for as long as possible. Even though the Recycle Fibre Plant had no speed as such, there was pressure to ensure quality was very high.

Why did you think it was safe?

Because we as a group of five shifts had evolved a system that was working.

What were your priorities?

To get the machine up and running as quickly possible, so we continued to feed in the bales to keep the towers full and ultimately the paper machine fed.

Understanding of management priorities:

I can only conjecture that their priorities were to ensure the Recycle Fibre Plant run as efficiently as possible.

Time pressure:

It doesn't matter what role you have, there is always a time pressure. As we had only just started the 100% run, the blockages put us on the back foot.

Why put yourself in that position?

At the time it wasn't dangerous, it was part of the job. I knew that when working with the returning bales there would be blockages. After all, why do people speed down motorways way over the 70mph limit? Because it 'feels' safe.

Everybody knew their roles that day. It wasn't like we hadn't done it before, we had been unblocking the thing for a whole year.

What pressures were you under? And what was your relationship with your boss?

As mentioned, the blockages happened almost as soon as we turned the machine on so we were acutely aware we needed to get it going again pronto.

I would say though that I had a good relationship with most of the management team, including my boss. I like to think I was seen as a valuable member of the group.

How to do it better:

The better way was what was put in place afterwards – adding an interlocking key system and maglocks on the hatches. Another thing that would have helped was to ensure no bales were on the left side of the incline conveyor.

What do you think the cause was?

It was simply a perfect storm. Everything lined up to explode. You can't just say it was one thing that caused the accident. I think the second blockage was the straw that broke the camel's back. We could afford a blockage of forty-five minutes at most, after that we were cutting it fine.

Experience:

I had nine years' experience at the mill. I worked in the Conversion Department for five years and spent the next four years in the Recycle Fibre Plant. I was well accustomed to working around the site and to the processes in place to get the job done.

Shock and Awe

I have one of the best jobs in Health and Safety and I'm lucky enough to travel the world talking to people about it. It's no ordinary job and the qualification to do this role is to suffer a life changing injury. In fact, all you need to be is the first person in the United Kingdom (as far as I know) to have their arm replanted above the elbow.

It was one of 600 major incidents in the paper industry during the year 2000. That is four short-haul plane loads.

It is interesting listening to other people talk about accidents at work (I sit on courses to learn new skills, but also to pick up new bits for my presentations). And as ever, when people start to talk about accidents the victim of the injury is usual called stupid, an idiot etc. This is without knowing the context of why they were in that situation in the first place.

It is agreed that most people don't go to work to kill themselves. How come 70,000 odd people injury themselves (some fatally) every year? It's simply down to variables on any given day, like my accident.

- Miscommunication due to poor radio signals
- No written procedures
- Poor leadership
- Trust in knowing the people you work with
- Complacency

Take any picture that shows an unsafe act or condition. It is a result of the resources available to the worker. If the correct tools or equipment aren't accessible, then the person will adapt or evolve the equipment and/or situation to get the job done.

The question that needs to be asked is what is the essence of any industry or organisation? Answer: to get the product or service from point A to point Z (i.e. widget to customer) with the minimal cost, but maximum profit. If workers are not supplied with the right tools to complete the task with the minimal effect and maximum results, then they will adapt or evolve to achieve the result.

Fig 14: What a view

So, how did I get into talking about my injury and the lead up to what was an avoidable accident? The answer **Good Fortune**, like most people at work, who survive a major accident. There needs to be that element of luck, where things that line up to allow you to survive to fight another day.

I had a Facebook message from an old friend who had returned back home to the north east. The message was that he had been on a course and this guy had used my video from 999 to illustrate the dangers of unsafe working and he wanted to talk to me about the accident.

This guy was called Dan Terry and we met up to discuss the

accident. I thought it was just to clarify some points, but no, Dan wanted me to actually talk to people in their workplace about the whole thing! I always wanted to get people to learn by my mistake by donating that pint of blood or £1 to the local air ambulance or even to get first aid trained.

To speak to people about the accident was another thing, as I had never even thought about doing that before. The first speech was a disaster. I stumbled over my words and I was called every name under the sun about my accident, but other than that I survived to live another day.

Years later I now look back at that day – and other challenging days since - with glee; each day has had its own learnings.

I take on the words that the 40th President of the United States (Ronald Reagan) said:

"We can't help everyone, but everyone can help someone".

He was right. We can't get everyone in the audience to learn and retain the information given out. But we can get everyone to take at least one slide away with them and from there you can guarantee that a good majority of the audience will remember the session.

In Wales that was put to the test, when I was talking to the organiser about the quote and why it was in the presentation at the time. She asked me if I'd ever tested the theory. To be honest I hadn't, but I knew that in normal training sessions the average delegate will retain about 5 to 10 percent of the information given.

After my presentation, my host and I walked the site talking to others about the process and the presentation. At the end of the conversations with my host and others around site told me which slides that they related to most and these related to most slides. With that information and knowledge my host was more than happy that the presentation worked and Ronald Reagan was right.

To help with an organisation's behavioural safety program, the presentation has posters to go along with it. The posters are of the slides with a strapline where possible.

This helps as it is great to get people to talk to employees about their life changing injuries or family tragedies. The shock and awe such stories give is great to wake employees up out of their complacent habits for a week or two if you're lucky. But something else is needed and that is why I give away the posters.

Even this is time limited, so now I have moved on to develop a set of benchmarking questions that employees fill in after the presentation. They were developed because after each session you get told that there was lots of feedback about the presentation and it gives delegates the courage to speak up about issues. But is all this information and communication gathered and recorded? Probably not. My point is, the session caused the shock and awe needed and it created a discussion point, but it is useless without the information and communication generated captured and recorded. Yes, urgent issues are addressed, even at midnight once when I did a night shift talk and the engineers cornered the safety manager to say that they had an issue similar to mine involving isolations and the remoteness of them to the operation they needed do.

Behavioural safety in the last year has started to change, especially with questions about the culture of the organisation and discussions around how employees develop systems quicker than the written processes can keep up with.

So, has the road come to the end for the shock and awe presenter? I would like to say no, but it needs to change and evolve. It is no good coming into an organisation and shocking them with the outcome of an accident. The shock and awe generated lasts a couple of weeks for most as they slide back into the old routine. It's like seeing images and stories on the news where people are killed or injured and their sub-conscious starts to put the barriers up to protect.

There needs to be take-aways for all concerned management who can track the success of the sessions. Plus, employees need to have something to remember from the session so they go home safe at the end of each day.

Why have I labelled myself and others that do what we do 'shock and awe speakers'? The answer is easy, because we are the shock factor in getting people to wake up to their habits at work. Many times, I have been told this and looking at people's faces as I play the edited 999 video, and they watch it and then look at me. You can see the shock all over their face as they calculate that the person who was standing in front of them talking about their workplace and behaviours, is the person in the video.

When we talk about shock and awe, most of us go back to the newsflash of March 2003 and the invasion of Iraq and Baghdad in flames as wave after wave of missiles hit Iraqi government buildings. The term shock and awe (technically known as rapid dominance) is a military doctrine, which is based on the use of devastating power and spectacular displays of strength. This is to blind the enemy's awareness of the battlefield and destroy the morale to fight.

I would not say safety speakers use devastating or spectacular military power or strength. What we do use is that devastating personal story and spectacular will to triumph over the injuries suffered. For myself I have a greater advantage as must people think I'm there to deliver some health and safety talk, as they can't see my badly scarred arm. Most of the damage is on the inside and it is my left arm. So, to them I look and act 'NORMAL' because we live in a right-handed world.

What else do I put my accomplishment down to, in getting the employees on side? Speaking their language; talking about their working environment and showing empathy about conditions and practices. This is helpful especially when I have had a site tour. I had feedback a few times when someone said "it was good to hear someone know what they are talking about!"

It's the same with the leaders of the business too. Talking about the latest business thinking about safety and strategy helps as business leaders are sponges to new thinking and adapting them to their environment. You really need to be a man for all seasons.

The thing that crosses all levels of an organisation is what I learnt from my days at Sainsbury's. Their customer training is all about the 'great hello', knowing your stuff. If you don't know the answer to something, be honest. Get the information, or find someone who does, and then give the 'great goodbye'.

Not everyone is won over by a smile and you trying to help them work a little safer. Studies show that 1% to 2% of the population are psychopaths and another 30% are narcissistic in nature. Like shock and awe, it doesn't have a 100% success rate in knocking the enemy out of the war.

How do you try and overcome these people in the group? There is no easy answers to this situation as everybody is different. I said earlier it's about talking their language, but this isn't always the case. I have developed the presentation to peak and trough, like a rollercoaster, taking their emotions on a ride.

Treating the slides like an onion helps as well. You keep peeling away the layers until you reach the level where the delegate will change. You make the sub-conscious to *want* to change. The last resort is to hope that peer pressure kicks in and the group convinces the hard core that they need to change or brutally refuse to work with them.

Shock and awe is relative to the person witnessing it. I recently attended the CITB Health and Safety Awareness course to get my CSCS card. It's always good to learn from others about how they teach and share ideas.

I'm pleased to say I achieved a 100% pass mark at the end of the course assessment. It was surprising to see a trend develop with the wrong answers given by the other candidates. There were three questions in all and these involved:

- Who you do report an issue to on site if another contractor is not working safely?

- What is risk?
- What is the biggest cause of fatalities in construction?

The first question had a mixed response amongst candidates. The other questions had more or less the same answers. The answer to risk was potential, this was mainly down to the tutor slides and explanation not being clear.

The last answer was the candidates thought it was electrocution! Even though it was highlighted throughout the course that falls from height is the biggest killer in construction. I believe people thought it was electrocution because that past of the course was the last part and it involved the shock and awe videos which certainly stuck in the mind.

Shock and awe is an important part of enhancing behaviours in the workplace. It does have to be remembered for the right reasons though and not as a soap opera that someone came in to work and spilled their feelings about. Employees then have no connection to the person in front of them, thinking "we don't do that thing in my workplace". It needs to be part of the bigger picture in ensuring people go home from work in the same good health as they arrived in.

We are a species that evolves as we learn and put the knowledge into place. Health and safety is evolving too, and the safety speakers who use shock and awe need to change and enhance their tactics as well to keep hitting the message home. The audience need to truly understand that if they get injured it isn't just them that take the hit, friends and family suffer too.

The Future of Health and Safety

The future of health and safety isn't going to change in the forthcoming years. It might change a bit to become more integral with human resources, but the actual role will stay the same. It will always ensure people go home safely every day. It also ensures that the business is keeping to its legal, financial and moral obligations to a healthy and safe workforce and environment.

As we start moving towards the end of the 21st century, and enter the 22nd century, we will see more and more of these changes taking place in the organisations we work in.

It should not be anything to be frightened of! It should be embraced and used to our advantage, just like the industrial revolution of the 18th century and the technology leap at the end of the 20th century.

Health and safety is currently in flux as it leaves regulations behind, as these are the foundations to any good practice. The walls and roof need to be put in place, but at the moment nobody really knows where to put the windows or doors in the walls.

Health and safety is a little like panning for gold. You stick the pan in the river bed and pan. Getting rid of the big stones and then washing down to the fine grit, to where little grains of gold appear.

These little grains of gold are the change and future of health and safety. There are going to be no more big changes in the evolution of safety.

The future is technological and will be based on computers and robots. You only have to watch the latest tech TV shows to see these changes taking place.

Health and safety will change when the present and future workforce changes from human to robot. At the moment robots are only automated, they are not self-aware. Robots can't currently do the dull, dangerous and dirty jobs where interaction with the outside world and imagination is needed to complete tasks. These jobs are still a human domain for now.

Once robots and the computers that control them start to become self-aware, they will take over the dull, dangerous and dirty jobs, for example warehouse, sewer and cleaning work.

With fewer humans to look after in the working environment, the role of the health and safety professional will evolve into something else. It will be more like a data analysis role, where information is collected from the robot about environmental conditions and its interactions with its surroundings.

To see a change of roles you only have to look at the modern day airline pilot. The role is more a supervisor of instruments of the flight deck and to give a reassuring presence of control for passengers. Yes, the pilot is hands on during take-off and landing, but modern planes can really fly themselves.

If the truth be known, the pilot's role is to correct and fly the plane in an emergency, because the human is still superior in reacting to a crisis situation.

Health and safety professionals of the future will be no different to the modern-day pilot now. They will just over see the instruments and react to any emergency situations that may arise.

Conclusion

To conclude this book please watch the following trailer from the film "Sully". It's a great film and the trailer explains everything in two minutes about this book and my story.

As Sully says he was a pilot for 40 years and in the end, he will be judged on 200 odd seconds! Most of us work for 40 – 50 years and in the end of a productive career, we have the potential to be remembered for just a few seconds for making an outstanding decision that solved an issue or the disaster that we were involved in. It is normally it boils down to a millisecond of a choice.

I thought I would leave you with one more story about how employees see health and safety professionals, and how health and safety professionals see employees.

A party of suppliers was being given a tour of a mental hospital.

One of the visitors had made some very insulting remarks about the patients.

After the tour, the visitors were introduced to various members of staff in the canteen.

The rude visitor chatted to one of the security staff, Bill, a kindly and wise ex-policeman.

"Are they all raving loonies in here then?" said the rude man.

"Only the ones who fail the test," said Bill.

"What's the test?" said the man.

"Well, we show them a bath full of water, a bucket, a jug and an egg-cup, and we ask them - what's the quickest way to empty the bath," said Bill.

"Oh, I see, simple - the normal ones know it's the bucket, right?"

"No actually," said Bill, "The normal ones say pull out the plug. Should I check when there's a bed free for you?"

For me, the safety world needs to get a bit more creative because that is how mankind evolved. We found our way out of the cave and have been creative ever since!

The safety world is seen to put the brakes on everything, that it stops us doing anything. This is not the case at all. Unfortunately, people have used the word "safety" for their own needs and requirements and in doing so have given health and safety a bad name. The safety world has actually helped them in this agenda by sitting back and not correcting the issue at hand to get things back on track.

Safety should go hand in hand with everything we do. It should be integral. It shouldn't be a bolt-on to keep solicitors and journalists at bay should things go wrong.

Loosen the chains and think outside the box! But be warned; as

Horowitz said about chess. 'One bad move nullifies forty good ones'! Remember, don't lose the chains, but loosen them!

Chapters - Circumstantial

Background

How did I get into speaking about my accident and inspiring others? I must thank Steve Foster and Dan Terry*. Firstly Steve, for attending a safety course at SCA Prudhoe and speaking to Dan about how he knew me, after Dan had shown my video as part of his course Working and Behaving Safely (WaBS).

Secondly, Dan for having the foresight to encourage me (and people like me Jason Anker, Paul Blanchard and Dylan Skelhorn) to speak out about the consequences of getting it wrong at work.

Dan must have questioned what on earth he'd done at first. I used to phone him constantly saying I had this idea or that idea for the presentation. This was all done normally, while I was down the beach walking the dog.

If anybody knows me truly they know I love bouncing ideas off walls and the more you throw the most will (hopefully) stick.

A big thank you is due to my industry (paper) via the *Confederation of Paper Industries* (*CPI)* for allowing me to talk at their conference in 2013 in front of the movers and shakers of the industry.

Some of these movers and shakers need a thank you too for giving me the chance to speak to their mills and factories.

**Dan Terry runs Ascend Training Consultancy Limited - www.ascend-consultancy.co.uk*

While in hospital, it soon became clear that I was no ordinary patient. On the Monday morning Mr Davidson came down to see me and to see how the arm was doing.

He squeezed my index finger (which I couldn't feel at all) to see if the circulation had improved. Luckily it was ok. Before he left he asked if I was willing to speak to the press or if the operation could be released.

I didn't have an issue with the news being released, as long as it was to the broadsheets (The Times, Guardian and Daily Telegraph). I wanted at that early stage for people to learn from my mistake and the ensuing operation. I didn't want it to be yesterday's chip paper.

Favourite Presentation

I really haven't got a favourite presentation. Each one is special because each location and audience is different.

During my time presenting, I have really enjoyed the travel. Okay, even luxurious hotels are never like staying at home, but I still loved it.

Travelling meant seeing different parts of this beautiful country (and world). I have seen things that I didn't ever think I would see during my working life. Yes you might see some things on holiday, but to actually be paid for it is seriously great!

Walking around sites and factories is fascinating too. It is, at times, like being on your very own real life 'How's It Made'. From being close-up to ships when working at docks (at home they are way out in the Thames Estuary and only look about an inch big) to seeing teabag and coffee filter paper made (busman's holiday that one).

My favourite walk was around a part of Bournemouth airport getting up close and personal with the planes. I even stood underneath one of the biggest private jets in the world. I'm a plane geek so that really made my day...

Recently, I had a personal tour in a van around Heathrow and a private viewing area overlooking one of the runways on a silo of a clients. It still amazes me how them Airbus A380s (double decker planes) ever take off?

I enjoy interactive presentations afterwards and the sharing of stories. You get a real sense of where others have been in the past. It's great to see that little bit of empathy in their eyes as I share my story.

The most amazing moment during a presentation was when a young apprentice lad with one arm tied his shoelace up using his hand and mouth! Really, he did it in about 16 seconds! To prove it wasn't a fluke he did it again, to everybody's amazement.

It is them moments like that which really do make my day, as it adds that little extra sparkle to the presentations. Not only do people talk about the presentation, but they talk about those moments too.

I'm going to finish this chapter with a quote by South African musician Vusi Mahlasela

> *There have been so many great moments, and I feel so blessed every day.*

And that I really cannot say anymore on the great journey I have been on and continue to do so.

Why Great Stories are Power Tools for Sharing Knowledge

Throughout time and history, we've been using stories to pass on wisdom and culture. Stories are indeed a great way to share knowledge. It all started once humans learnt to light a fire and use language (now it is called watching the goggle box). By telling a story, however tall, we can communicate lessons, convey complex concepts, or represent abstract ideas.

So it's not surprising that some of the world's leading organisations use stories as a tool to not only educate their employees, but also to motivate and inspire them.

Stories exist within all organisations. It's common for individuals to share their experiences and goals through anecdotes or narratives over lunch, during meetings, or at the water cooler.

Collectively, stories can do a lot for the workplace. They include the following:

- **Strengthen the culture** – stories can help to convey standards and values derived from the organisations' past and also to build its future. This enables people to connect around a shared narrative that enables employees to believe in the direction.
- **Build trust** – stories can convey information about the

organisation's trustworthiness to its employees and to its customers and suppliers.

- **Transfer tacit knowledge** – stories about work and real-world situations help to convey tacit knowledge which is usually embedded in formal processes (policies, risk assessments and safe systems of work) – in a form that is easier to understand and convey to others.
- **Facilitate change** – Stories carry an emotional element that rational arguments lack.
- **Help employees remember key concepts** – Stories have always had the ability to engage our emotions through the connections they make. They make us feel something and it is this emotional response that makes knowledge the glue.

What makes a great story to remember?

Stories for knowledge sharing prioritise informing over entertaining, but it has got to hold people's attention. They serve to convey both information and emotion, to communicate both tacit and explicit information, and to provide the context for the vision and execution of it. For a story to be a good knowledge-sharing one there are two main ingredients:

#1 Simple

A good knowledge-sharing story can be easily communicated in the flow of work. It's important that it's designed to make specific points that you want to get across. This is done by cutting out all the fluff and avoiding too much detail that might distract from the main idea, or steer away from the impact you want.

#2 Relatable

The story has to be relevant (or can be drawn in comparisons) to the activities and concerns of its audience. Otherwise it will lose the impact you want to achieve. Even though the audience might not have directly experienced the situation in the story, it must be probable that they could experience something similar in the course of their work or life outside.

Opportunities to infuse stories into the workplace

There are some key moments that present great opportunities for us to share stories and knowledge. They include:

#1 Kick-starting a new idea

Suppose you're launching a new product. A great way to get people excited and involved from the start is by kicking it off with the story behind the conception of it. It provides the direction and vision of the launch. The compact nature of stories makes them easy for people to retell. This in turn reaches a greater audience in getting them to support the new idea, as they feel they have ownership.

#2 Socialising new employees

Stories come in handy when trying to share culture and norms to groups of new employees. Examples of stories to share could include: founder stories to set the tone for resilience and innovation, or success stories to motivate, inspire and communicate expectations. These could include how the organisation grew to the success it is today.

#3 When leaders impart wisdom

Stories can vividly communicate a leader's wisdom. Leaders can use stories to share their experiences, insights, best practices and know-how in a manner that leaves a lasting impression on listeners.

#4 Presentations needing impact

During a presentation, when the audience are starting to switch off and fidget. A good short story breaks up the presentation and reawakens the audience and the subject being addressed.

The Operation

How the hell do you start with this?

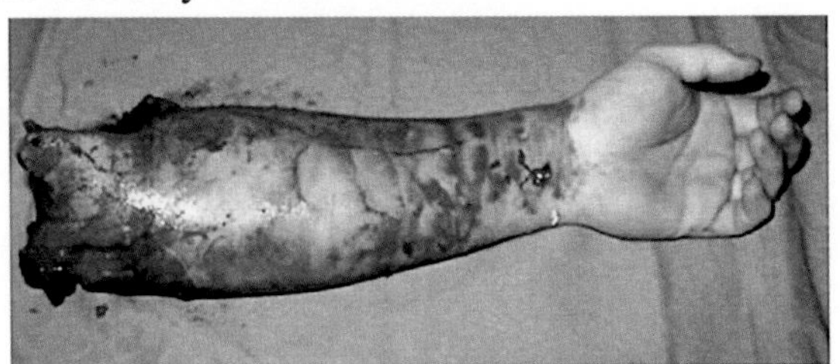

And end with this?

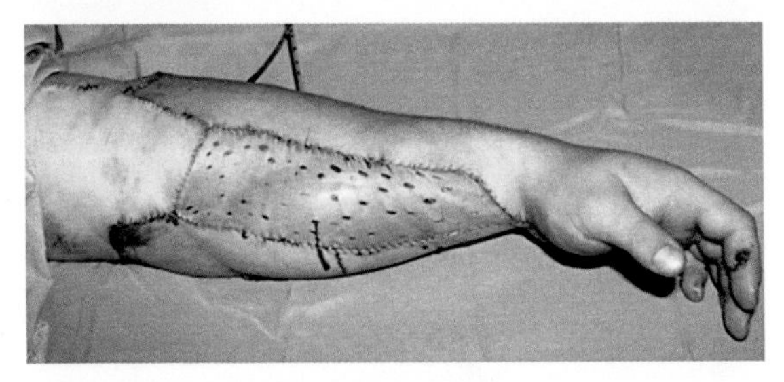

...a (nearly) fully functioning limb.

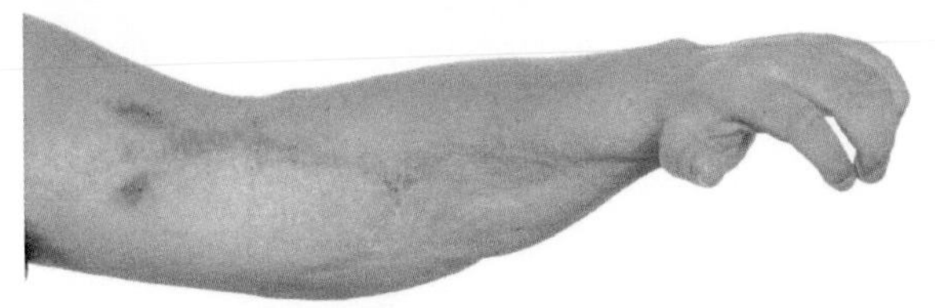

Well, I can't tell you. On my part it was mind over matter to get it going again.

What should someone do in the event of an accidental amputation?

Doctors advise that the limb should be put into a plastic bag and sealed before being placed into a bucket of ice and water. The part should not be placed in direct contact with ice.

How long can a limb survive if it's kept cool?

A finger can be reattached up to 8 hours after amputation. An arm is unlikely to be successfully reattached if more than 3-4 hours pass.

What are the chances of a successful reattachment?

If it's a clean cut and the patient is otherwise well the chances of being able to replant are very good using microsurgery.

How is a limb reattached?

Initially the bone is reattached. Surgeons then generally attempt to reconnect the blood vessels to get blood flowing again. After this the nerves are reconnected. The whole procedure takes many hours, often lasting an entire day.

How common is limb reattachment?

Full arm reattachment is rare – most of the 49 plastic surgery centres in the UK will only encounter this every two years or so. Operations to reattach thumbs or fingers are more common, taking place approximately once a month at each centre.

How long does recovery take?

Generally, patients will successfully regain movement in the reattached limb. The difficulty is in getting the nerves to work again and restore feeling. New nerves must grow from the old end of the cut nerve and this can take many months, particularly for older patients. Some may never fully recover feeling in the severed limb.

How long have surgeons been reattaching limbs?

Microsurgery – the key skill for reattachment – has been around since the 1970s but it didn't become mainstream until the 1980s.

Sources: Dr Hamish Laing, Reconstructive Plastic Surgeon.

British Association of Plastic Reconstructive and Aesthetic Surgeons.

Post Traumatic

In my chapter “Coping” I feel I gave a brief outline of how I coped with what happened to me.

Sixteen years on, organisations are more knowledgeable about mental health amongst employees and contractors. There needs to be warning here though: Where the majority of cases of mental illness is real, there are some that will jump on the bandwagon, just like the few who fake back pain and other ailments.

Life is a rollercoaster and we all face ups and downs. On the whole life is on the level and steady. I read a couple of years ago about an interesting model called Slight Edge by Jeff Olson. It’s a great model to show the difference between failure and success.

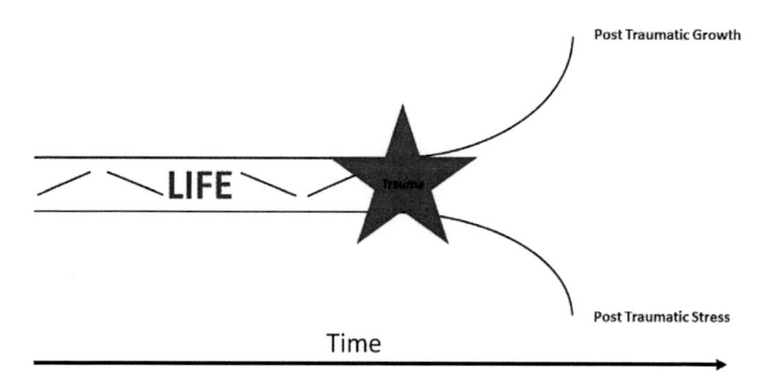

Fig 15: Recovering from Trauma

The above model shows what the possibilities are to people when they suffer a trauma. At the moment, most people think that if something horrible happens then they will suffer post-traumatic stress and the experience will forever be a blot on themselves.

I remember reading the counsellor's report when I was assessed for my mental state during the legal process. It said that I didn't suffer from PTSD from the accident. Torie always disagreed because she lived with me at the time. If anything, it was frustration because I couldn't get things done like I used to.

I have always been an optimist; a glass half full type of person. I guess that's why I've bounced back after disappointments.

When I look back at that time on the floor with Darren, there were times when it was quiet and I had a chance to reflect. Perhaps I could go back to college and pick up where I left off at the age of seventeen. A second bite of the cherry if you will. I started to look forward.

In hospital, there were never any goals set. I set ones for myself though, like going to the bathroom on my own and writing with my other hand.

One thing I always wanted to achieve was to get people to learn from my mistake; to learn first aid, give blood and donate to the air ambulance services around the country.

What I have just talked about in the last few paragraphs is a thing called Post Traumatic Growth. So what is it?

Post-traumatic growth (or PTG) involve finding positive psychological change as the result of adversity and other challenges. This allows you to then reach a higher level of functioning.

Others would say it's a motivated positive illusion to help people get over the trauma. It protects them from psychological damage caused by the incident.

The memories of that day will never leave me, from the moment of the crack of bone to closing my eyes in the operating theatre. I will remember that time forever. So, how have I kept the demons away? My answer is to turn them into fond memories!

People sometimes say to me "machinery always win you know!" At this point I smile and say "Well, on the 25th November 2000, I managed to score a draw!"

I tell people I was in the wrong place at the wrong time, but luckily everything else was in the right place at the right time!

Mental health is about making sure we understand what makes people tick. We need to encourage an open culture where people feel free to express their feelings. The atmosphere of opportunity must always be present.

The Shoe Story

It was reported many years ago, that two salesmen were sent by a British shoe manufacturer to several African countries to investigate and report back on market potential for selling their latest designs.

The first salesman reported, "There is no potential here - nobody wears shoes. Don't send any samples."

The second salesman reported back, "There is massive potential here - nobody wears shoes. Send samples now!"

We need to remember the following:

The optimist says the glass is half full.

The pessimist says the glass is half empty.

The project manager says the glass is twice as big as it needs to be.

The realist says the glass contains half the required amount of liquid for it to overflow.

And the cynic... wonders who drank the other half.

Thank You

My partner in crime Cheryl deserves a big mention. She has given me the encouragement to follow this unique journey to inspire others to a safer life.

I was a baker when we first met, a steady job in Sainsbury's paying regular wages and hours. Then, this fantastic offer came along to do something out of this world. I'm sure she thought I was mad, actually I'm sure she still does when I talk about new slides and ideas.

As the bookings to speak grew and the hours at work decreased, she finally encouraged me over a beer on holiday in Corralejo, Fuerteventura to make my mind up! So, after 5 years at Sainsbury's I went full time as a speaker, trainer and safety professional.

Cheryl is not just my partner, she can be my biggest critic - in a good way. She keeps my feet on the ground and helps me see things from a new perspective.

I would not be human if I wasn't to leave the biggest thank you to the man who not only saved me, but my arm too! Mr Darren Carpenter Esq.

We go to work to provide for our families, but to go to work and save someone's life is a whole other thing! For an hour and a half, he stuck his fist in my arm pit and kept me going. It must have been made even harder when I asked him to tell my family I love them in case I die. Darren kissed my forehead and said something like "you're going to tell them, I'm not!"

Darren kept talking to me to keep my spirits up – and to keep me awake. If I was going to die that day I wasn't going to go out with a whimper, I was going out with a laugh. The Black Knight sketch from the Monty Python's Holy Grail was the inspiration for a smile or two between us.

Many times I've been told I was brave and a hero. I may have been brave, but not a hero! The hero of the day - and since - has always been Darren. The bloke was out of this world that day to do what he did! Thank you, Darren.

I shouldn't forget to mention the medical teams, paramedics and the police. All those who put me back together that day to get me where I am now. The truly amazing thing is these people do it day in, day out.

I was asked once about the NHS and I was very frank in what I said: "The NHS is world class when it comes to pushing the boundaries of surgery and care".

I would love to thank my Dad too but sadly that isn't possible – as I mentioned, he died when I was sixteen. But the best thing he ever taught me was to be honest and always be enthusiastic no matter what you do.

Mum, like all mums, is my rock. She has told me some tales of what happened over the first 48 hours after my accident. The best one was when the medical team explained all the options available and Mum in her frank mode said, "there's always the what if!"

Thanks Mum!

Acknowledgements

There are a few people I need to acknowledge on this journey over the last few years.

Dan Terry for the encouragement to explore and challenge myself to reach a different level. I'm sure can tell you a tale or to about the crazy ideas I had while walking the dog down the beach and talking to him as he drove to a client or too.

Colin Murray a great guy to work with when training and have a pint or two of the local ale, when we are out and about.

John Mackay if you ever needed a mentor. John is the man to go to, to discuss the world and putting the world in a different context

Professor Richard Booth the loveliest and gentlest man you could meet. Many a day or night we have discussed things from ancient history to modern day thinking and the current situation with F1.

Finally, little sister Faye for being my little sister.

Closing Thoughts

'We can't help everybody, but everybody can help someone'.
- Ronald Reagan

It's true we cannot help everybody because we haven't the resources or time. But everybody can help someone. If at some point we all just stop and help one person, in the end everyone is helped eventually.

I was going to the United States (Buffalo in fact) and I was being sensible, as I was going to a conference and having a Coke Zero. the can, I had noticed, was this… LEGEND…

Please remember. No matter what you do in life, you are a legend!

Who Am I

Born Christmas 1972 in South London, I am the eldest of two. Dad was a Brickie for Higgs and Hills and Mum was a dressmaker for a little Jewish family business in Balham.

I spent the first couple of years living in a flat in Tooting. Tooting has since been made famous by Citizen 'Wolfie' Smith (Power to the people) and the present Mayor of London Sadiq Khan.

In late 1974 Mum and Dad were given the chance to move to Sittingbourne in Kent. I went to South Avenue infant and junior school and St John's secondary school (now Sittingbourne Community College). My GCSEs were a disaster as my Dad died on the 18th May, right in the middle of my exams. It was a bittersweet fortnight, as Chelsea - his club - won the Second Division (now the Championship) and the Arsenal (my team) beat Liverpool 2-0 with the last kick of the game to win the First Division (now the Premier League) title.

After school, I went to Thanet College (now East Kent College) and studied for a diploma in Business. My ambition was to be a bookie (turf accountant) or a pub landlord. Not your normal ambition for a sixteen year old!

I left college and got a job at Kemsley Mill in Conversion, working my way through the ranks until I decided to push the boundaries and joined the Recycled Fibre Plant.

In 1998 my daughter Georgia was born and I married my first wife, Nick. It was a happy marriage, even though Georgia was born with issues (cerebral palsy and epilepsy). The marriage collapsed in 2001 after the strain of Georgia's illnesses and my accident.

In 2002 I met Torie, my second wife. She had a daughter, Mollie. The year after meeting I had another operation as the plates in my arm were keeping the fracture open and the bones were not healing (broken bones need to rub together to fuse and heal).

For nine months, I wore an external frame. I had every left arm cut out of my polo shirts and a few jumpers. I even got stuck to the odd door frame when I forgot I was wearing several pounds of metal. The frame was taken off the Friday before England winning the Rugby World Cup, so that weekend it was a double celebration.

I returned to work again in 2004 and worked on Safe/Standard Operating Systems (SOP). It was not very well brought in by some managers and staff. After constantly banging my head against the wall, I recommended some changes were needed. Unfortunately, my direct manager and I didn't see eye to eye about the buy in and new direction. It was best all round that I left.

For four years I was a house husband and looked after Mollie. This allowed Torie to progress her career in the NHS. With Mollie getting bigger, I joined Sainsbury's as a trainee baker.

In 2010 my six-year marriage to Torie broke down. It was dissolved in 2011, after nine years together.

In September 2011, I met Cheryl. She used to introduce me as her toy-boy to her friends although six years later I'm actually now her carer, as she jokes as we get older.

A year later I met Dan Terry and as they say the rest is history. I have travelled the world, spoke to thousands of people and seen some brilliant things made and built.

And now, against the odds, I've written a book. Who'd have thought it…

References

Paul O'Neill opening speech to Alcoa
http://www.huffingtonpost.com/charles-duhigg/the-power-of-habit_b_1304550.html

Medical Photographs from Queen Victoria Hospital
East Grinstead

Stock Pictures and Photographs from Shutterstock

Video of my accident from BBC programme 999

Sunday Roast Story
http://selfdefinedleadership.com/blog/?p=158

Videos from Youtube.com

Loophole Story
https://en.wikipedia.org/wiki/Loophole_(short_story)

Alexander meets the Yogi
https://andreaskluth.org/2010/03/12/alexander-meets-a-yogi-whos-the-hero/

Airline repair
http://www.dailymail.co.uk/travel/travel_news/article-3102312/Shocked-passenger-takes-photo-airport-worker-using-TAPE-engine-shell-easyJet-plane-moments-off.html

Sparrow Story
http://texafied.com/blog/2009/03/28/like-a-sparrow-flying-through-a-mead-hall/

Figure 6
http://lackeby.com/screw-conveyors/

Figure 7
http://www.kwsmfg.com/resources/problem-solvers/food-grade-screw-conveyor-cookie-toppings/

Bishop and the missing ladle
http://www.businessballs.com/stories.htm#the-bishop-priest-ladle-story

Drowning Man
http://truthbook.com/stories/funny-god/the-drowning-man

Houdini
https://douglasvermeeren.wordpress.com/2011/10/09/houdini%E2%80%99s-great-mind-trick-was-on-himself/

Figure 8 & 10: Warehouse
http://www.ftcsouth.co.nz/gallery/

Figure 9 & 11: Dock
http://www.safetycareblog.com/2011/05/unsafe-work-practice.html

Transactive Memory
https://en.wikipedia.org/wiki/Transactive_memory

Rogue Monkey Story
http://www.businessballs.com/stories.htm#monkey

Figure 12: Many hands make light work
http://prevenblog.com/cuando-un-accidente-en-el-trabajo-deja-de-ser-accidente-de-trabajo/

The Philosopher and the King Story
http://www.uexpress.com/tell-me-a-story/2013/3/24/the-philosopher-and-the-king-a

Hunter and Priest Story
https://ia601408.us.archive.org/6/items/kottobeingjapane00hearuoft/kottobeingjapane00hearuoft.pdf

President and the Janitor
http://thelegacyofyou.com/kennedy-the-janitor-putting-a-man-on-the-moon/

Falling Trees
http://www.inspirationalstories.com/1/119.html

Teamwork Story
http://fortune.com/2011/11/11/steve-jobs-the-parable-of-the-stones/

Marmite and Marmalade remark
http://www.bbc.co.uk/news/uk-politics-33687369

The Wolf Pack
https://leandroherrero.com/the-wolf-pack-and-leadership-the-lessons/

Butterfly Story
http://www.businessballs.com/stories.htm#the_butterfly_story

When the winds blow
http://www.knowledge-share.in/moral-stories-4/#.WWnpv4jysak

Maybe
http://www.drmarlo.com/?page_id=181

Society of Estate Clerks of Works
http://quoteinvestigator.com/2017/03/06/tap/
Starfish Story

https://eventsforchange.wordpress.com/2011/06/05/the-starfish-story-one-step-towards-changing-the-world/

HSE Statistics
http://www.hse.gov.uk/Statistics/)

Figure 13: What a view
http://www.theconstructionindex.co.uk/news/view/carefree-scaffolders-caught-on-camera

Mental Hospital Visit Story
http://www.businessballs.com/stories.htm#bath_bucket_story

Inspired to Sharing Knowledge
http://www.epiphanyedu.com/blog/2015/10/04/how-great-stories-are-powerful-tools-for-sharing-knowledge/

The Shoe Story
http://www.businessballs.com/glass-half-full-empty.htm

Stopping the Ripple

I saved a life that day, I chose to speak up.
It wasn't that I was brave; Nor had the time, I was just there.

I didn't want to be seen the hero, Or the grass about a safety rule. I knew they had done the job before. It was time to make a stand.

The hazards didn't seem that bad. I've done the same in the past; we both knew I had.
But today, I wasn't walking on by. The risk just seemed different today.

They didn't take the chance; As I spoke up. And with that act, they live another day.
I saved a life that day, as I chose to speak up.

Now every time I see them, I'll know I saved their life. And mine.
I have no guilt to bear, but it's something I need to share.

If you see a hazard that others take, that puts them at stake,
Be proud to question, that helps them live another day.

If you see a hazard and don't walk away, then have pride to say:
I saved a life today, as I chose not to look the other way.

Inspired by Don Merrill's I Chose To Look The Other Way

The Mahoney Family motto is:

We defend ourselves and our faith

900 years later the first part is still true we still defend ourselves, but I think it is time to evolve and change the last part to what we believe in.

It is important to defend what you believe in (not just your faith) as how can we challenge and move forward, learn and evolve if you don't have difficult conversations to this modern world.

If you would like more inspiration.

Connect on Linkedin
Paul Mahoney Tech IOSH

Go to the websites:
www.inspiringsafety.org.uk
www.pauljmahoney.co.uk